# Grounding Electrical Distribution Systems

# RIVER PUBLISHERS SERIES IN POWER

*Series Editors*

**MASSIMO MITOLO**
*Irvine Valley College*
*USA*

The "River Publishers Series in Power" is a series of comprehensive academic and professional books focussing on the theory and applications behind power generation and distribution. The series features content on energy engineering, systems and development of electrical power, looking specifically at current technology and applications.

The series serves to be a reference for academics, researchers, managers, engineers, and other professionals in related matters with power generation and distribution.

Topics covered in the series include, but are not limited to:

- Power generation;
- Energy services;
- Electrical power systems;
- Photovoltaics;
- Power distribution systems;
- Energy distribution engineering;
- Smart grid;
- Transmission line development.

For a list of other books in this series, visit www.riverpublishers.com

# Grounding Electrical Distribution Systems

**Gregory P. Bierals**

Electrical Design Institute, USA

River Publishers

Routledge
Taylor & Francis Group

LONDON AND NEW YORK

**Published 2021 by River Publishers**
River Publishers
Alsbjergvej 10, 9260 Gistrup, Denmark
www.riverpublishers.com

**Distributed exclusively by Routledge**
4 Park Square, Milton Park, Abingdon, Oxon OX14 4RN
605 Third Avenue, New York, NY 10017, USA

**Library of Congress Cataloging-in-Publication Data**

*Grounding Electrical Distribution Systems / Gregory P. Bierals.*

Routledge is an imprint of the Taylor & Francis Group, an informa business

ISBN 978-87-7022-617-2 (print)
ISBN 978-87-7022-616-5 (online)
ISBN 978-1-003-20730-6 (ebook master)

NEC®, NFPA 70®, NFPA 70B®, NFPA 70E® and National Electrical Code® are registered trademarks of the National Fire Protection Association.

The IEEE Standards referenced in this book are from the Institute of Electrical and Electronics Engineers

While every effort is made to provide dependable information, the publisher, authors, and editors cannot be held responsible for any errors or omissions.

# Contents

# Introduction

To begin, we will identify and analyze the definitions of terms in Article 100 of the NEC that relate to Bonding and Grounding. These terms and definitions must be thoroughly understood before going further into our study.

And, before we address the proper bonding and grounding methods, we will identify the AC Systems that are <u>not required</u> to be grounded, as well as those that are <u>not permitted</u> to be grounded.

In addition, Impedance Grounded Systems will be identified and treated in a similar way as ungrounded systems, although there are added benefits in the use of these systems.

It has been my experience that the topics of grounding systems and equipment are some of the most widely misunderstood subjects in the electrical industry.

There have been so many books and publications relating to proper bonding and grounding methods. And yet, there is still a significant amount of misunderstanding. And so, this book is designed to clarify these issues.

Throughout this book there are references to the appropriate Sections of the National Electrical Code. This has been done to further clarify my intention of stressing the purpose of the NEC, that is, the practical safeguarding of persons and property from hazards arising from the use of electricity (Section 90.1(A)).

It is my sincere desire to be of help in the proper application of installation methods that will assure safe and reliable electrical installations.

Of course, personal safety is of paramount importance, and this can never be compromised. But the proper means to install electrical systems that enhance the reliability and proper operation of electrical equipment is of equal importance. These two methods of protection go hand-in-hand, and neither may be overlooked.

As always, I welcome your comments and suggestions in order to better serve the electrical industry.

*Sincerely,*
*Gregory P. Bierals*
*Electrical Design Institute*

# 1

# Grounded and Ungrounded Systems
# System Grounding

System grounding is sometimes a requirement, especially when dealing with the most common types of premises wiring systems. For example, Section 250.20(A) and (B) relate to AC systems that <u>must</u> be grounded.

First, for systems operating at less than 50 volts, the following conditions apply:

1. If the supply transformer exceeds 150 volts-to-ground
2. If the supply transformer is ungrounded
3. Where the circuit operating at less than 50 volts is installed outside as overhead conductors

In the first case, if the primary of the transformer exceeds 150 volts-to-ground, there is a possibility that a primary-to-secondary fault would elevate the secondary voltage to an unsafe level, which may be a hazard to people, as well as to equipment. The secondary ground (earth) connection will limit this voltage-rise.

This also applies to the second provision where the transformer secondary, operating at less than 50 volts, would have a means of limiting any voltage-rise above ground (earth) potential, due to the fact that the primary circuit has no ground reference ('0' volts).

The third condition relates to the possibility of the exposed outside overhead conductors making contact with higher voltage lines, or lightning induced influences.

For AC systems of 50-1000 volts, the following conditions apply:

1. If the system is grounded and the maximum voltage-to-ground is limited to 150 volts, such as the typical 240/120 volt single-phase system or a 208/120 volt, three-phase, Wye connected system.

   A 240 volt, three-phase Delta connected system may be operated as a corner-grounded system, where one-phase is intentionally grounded, and the other 2 phases operate at 240 volts-to-ground. This ground connection is optional, and not a requirement. But, such a connection

1

to ground provides protection from lightning, external power faults, and serves to stabilize the system voltage to ground.

2. Where a system is 3-phase, 4-wire, Wye connected and the grounded (neutral) conductor is used for neutral connected loads.

   This is the common 208/120, 380/220, 480/277, and 600/347 volt system. Once again, in order to stabilize the system voltage, especially where the neutral connected load is unbalanced from time to time, the ground (earth) connection must be made from the neutral point.

3. Where a system is 3-phase, 4-wire, Delta connected and the midpoint of one phase winding is used as a circuit conductor.

This is the 'High-Leg Delta System', where the midpoint of Phases A and C is grounded, and a neutral conductor is derived from this midpoint. The system voltage is a nominal 240 volts from Phases A and C, and 120 volts from Phases A and C to the neutral.

However, the voltage from Phase B to the neutral is 208 volts (120 × $\sqrt{3}$ (1.732)). Section 230.56 requires Phase B to be identified with an outer finish that is orange, or other effective means at each connection or termination. This applies to Service Conductors. Section 110.15 requires the same identification for other systems, such as, branch-circuits and feeders. For switchboards, switchgear, and panelboards, 408.3(E)(1) requires the 'High Leg' to be the 'B' phase (middle), and 408.3(F)(1) requires that this equipment be marked in the field as follows:

'Caution _____ Phase Has _____ Volts-to-Ground'

It should be noted that some utilities use the 'C' Phase as the High-Leg at the metering equipment because the meter requires 120 volts from the 'B' Phase for its operation. However, within equipment beyond the meter, the High-Leg must revert to the 'B' Phase.

Section 409.102(B) requires the same means of identification for Industrial Control Panels.

## Direct-Current Systems

Two-wire, Direct Current Systems operating at over 60 volts, and not greater than 300 volts, and that supply <u>premises wiring</u> are required to be grounded. There are exceptions, such as a rectifier-derived DC system that complies with 250.20. For example, where the AC supply for the rectifier operates at over 150 volts to ground (250.20(B)(1)), the DC system is not required to be grounded (250.162(A)).

Three-wire, Direct Current Systems that supply premises wiring are required to be grounded (250.162(B)).

The connection to the grounding electrode (system) is to be made at the source, or at the first disconnecting means or overcurrent device. There is a third provision that accomplishes the same protection through the use of equipment listed and identified for the use (250.164(B)).

The grounding electrode (system) for the DC system may be a ground rod, pipe, or plate electrode, (250.52(A),(5),(7)), or a concrete-encased electrode, (250.52(A,(3)). or a ground ring (250.52(A,(4)).

These grounding electrodes must be bonded to any other grounding electrodes that are associated with the premises wiring system (250.50, 250.58).

## Ungrounded or Impedance Grounded Systems

Section 250.21 covers the AC Systems of 50-1000 volts that are not required to be grounded. These are examples where it may be an advantage to operate as an ungrounded system due to certain safety or operational concerns.

These ungrounded systems operate without a solid conductor connection to ground (earth).

Examples from 250.21 include the electrical supply systems that are only used to supply industrial electric furnaces. Or, separately-derived systems used only for rectifiers that supply adjustable-speed industrial drives. Also for separately-derived systems that are supplied by transformers which have a primary voltage of 1000 or less, and are used only for control circuits, and that have conditions of maintenance and supervision that assure that only qualified persons service the installation, and continuity of control power is required.

Where systems are considered ungrounded or impedance grounded, there is normally a need to maintain continuity of electric service due to a hazard to people or to a process. Without a solid connection to ground and a grounded conductor extended to the source, a single ground-fault will not cause an overcurrent device to clear. It is probably true that an ungrounded system, with conditions of maintenance and supervision that assure that qualified persons service the installation, and a proper ground detection system, which is required by Section 250.20 (B), is as safe as a grounded system.

However, the protection from high transient voltages from lightning, power surges from the serving utility, and supply transformer faults, are compromised.

There will be a grounding electrode, which is connected to the equipment grounding system in order to reduce the effects of a voltage-rise above earth potential on equipment.

Where a system is ungrounded, Section 250.21(C) requires that this condition must be identified at the system source, or at the first disconnecting means. This requirement correlates with Section 408.3 (F)(2) for switchboards, switchgear, and panelboards. The marking must be suitable for the environment, legible, and permanent, and state, "Caution: Ungrounded System Operating at _____ Volts Between Conductors".

It should be noted that High Impedance Grounded Neutral AC Systems (250.36) must be marked as follows: Caution: High impedance Grounded Neutral AC System Operating at _____ Volts Between Conductors and May Operate at _____ Volts to Ground for Indefinite Periods Under Fault Conditions (408.3(F)(3)).

For Resistance Grounded DC, systems the required marking is as follows: 'Caution: DC System Operating at _____ Volts Between Conductors and May Operate at _____ Volts to Ground for Indefinite Periods Under Fault Conditions'. (408.3)(F)(5)).

Section 110.21(B) addresses: 'Field-Applied Hazard Markings'. And ANSI Z535.4-2011 provides information on product safety signs and labels. Similar markings apply to ungrounded DC systems (408.3(F)(4)).

The definition of the term 'Voltage-to-Ground' in Article 100 for ungrounded circuits is the greatest voltage between the given conductor and any other conductor of the circuit. A 480 volt ungrounded Delta or Wye connected system has a voltage-to-ground of 480 volts, according to this definition.

Section 250.22 identifies the circuits that are <u>not</u> to be grounded. These include:

1. Circuits for electric cranes operating over combustible fibers in Class III Locations (503.155). The ungrounded circuits may reduce the possibility that an arcing ground-fault would provide a source of ignition of easily ignitable fibers or flyings. However, conducting paths through moisture laden fibers or flyings may permit sufficient current to flow from bare conductors to ground. If the system is grounded, this condition may be a source of a fire. With the system operating as ungrounded, a fire, although a possibility, is not likely to occur.

2. Circuits in health care facilities as provided in 517.61 and 517.160. These conditions, at least in the United States, are from a time when flammable anesthetics were in use. This is not the case now, although in hospitals at several locations in Africa where I have worked, this is common. Flammable anesthetics include: cyclopropane, ethyl ether, etheylene, ethyl chloride, and divinyl ether. These are Class I, Group C

materials (500.6(A)(3)). And, 517.60(A) states that in a location where flammable anesthetics are used, the entire area is considered to be a Class I, Division 1 location to a level of 5 feet (1.52 meters) above the floor. The space above this height is considered to be above a hazardous location.

In recognition of the possibility of arcing ground-faults as sources of ignition, power circuits within, or partially within, a flammable anesthetizing location are supplied from an isolated power system. An isolation transformer, with a secondary voltage of not more than 600 volts between conductors and proper overcurrent protection may be used. The secondary of the isolation transformer is ungrounded (517.160(A)(2)).

Batteries or motor generator sets may also be used as the isolated power system. These systems must also have proper overcurrent protection, and they are ungrounded (517.160(A)(2)).

3. Circuits for equipment within an electrolytic cell working zone, as provided in Article 668. Section 668.20 and 21 prohibit portable electrical equipment used within the cell line working zone from being grounded. The circuits that supply power to receptacles for hand-held, cord-connected equipment are required to be isolated from other systems supplying other areas, and they must be ungrounded. The power for these circuits is to be derived from isolating transformers with a primary voltage not in excess of 1000 volts, and with proper overcurrent protection. The secondary voltage is not to exceed 300 volts between conductors. The secondary system is ungrounded and it is provided with proper overcurrent protection.

4. The secondary circuits of lighting systems as provided in 411.6(A). These lighting systems are supplied from Class 2 power supplies (725.2, 725.30, 725.121, 725.124, 725.130, 725.139 and 725.179). The maximum voltage is 30 volts. These lighting systems are listed as a complete system (411.4(A)). The secondary circuit is supplied by an isolation transformer, which is ungrounded (411.6 (A)(B)).

5. Secondary circuits of lighting systems as provided in 680.23(A)(2)).

The transformers and power supplies for underwater luminaires. These luminaires are installed below the maximum water level of a swimming pool, and must be supplied by an isolating transformer which has a grounded metal barrier between the primary and secondary, or an approved double insulation system. The secondary is ungrounded and the transformer and power supplies are to be listed for swimming pool and spa use.

High-Impedance grounded neutral systems are covered in Section 250.36 for systems of 480 to 1000 volts. These systems require qualified maintenance and supervision, as well as a ground detection system. Line-to-neutral loads are not supplied. For higher voltages, Section 250.187 would apply. Where these systems are 3-Phase, Wye-connected, the grounding impedance (resistor) is placed in series with the grounding electrode conductor and the system neutral point (the center point of the Wye). There will be a grounded system conductor from the neutral point of the transformer or generator to the grounding impedance. This is a fully insulated conductor and must be sized in accordance with the maximum current rating of the grounding impedance, but no smaller than No. 8 AWG copper (8.37 mm$^2$) or No. 6 AWG aluminum or copper-clad aluminum (13.30 mm$^2$).

The grounding impedance is sized in accordance with the normal capacitive-charging current, or slightly above this current. The capacitive-charging current will flow through the equipment grounding conductor, which is run with the ungrounded conductors to the grounded side of the grounding impedance and through the resistor to the neutral point of the transformer or generator.

For example, if the system is 480 volts, and the capacitive-charging current is 5 amperes, the ohmic rating of the resistor will be 55.4 ohms.

$$\frac{277\,Volts}{5\,Amperes} = 55.4\,ohms$$

The grounding electrode conductor connection to the grounded side of the grounding impedance is sized in accordance with Section 250.66, and is based on the cross-sectional area of the ungrounded service conductors, or the ungrounded conductors of a separately-derived system (250.32(E)).

If the grounding electrode conductor is connected at the first system disconnecting means, it must be at least No. 8 AWG copper (8.37mm$^2$) or No.6 AWG aluminum or copper-clad aluminum (13.30 mm$^2$). These conductors are required to be insulated (250.36(B)).

If the supply system is Delta-connected, the impedance device may be a zigzag grounding autotransformer. This method of impedance grounding may be used to detect a ground-fault on an ungrounded Delta system. In this case, a resistor is connected in series with the neutral point of the grounding transformer and ground. Once again, the ohmic rating of the resistor is in recognition of the capacitive–charging current of the system.

For a 240 volt ungrounded Delta system, a zigzag grounding autotransformer may be used to supply 120 volt loads by extending a neutral conductor from the grounding transformer neutral point. For systems of over 1000 volts, Section 250.182 applies to Derived Neutral Systems.

Another possible benefit from the use of this equipment is the attenuation of triplen harmonic currents, where the grounding autotransformer is installed near the nonlinear loads that produce large triplen harmonic currents. The transformer windings mitigate the harmonic currents in such a way as to prevent them from flowing upstream where they may cause overheating of neutral conductors and the supply transformer.

However, on premises wring systems, the use of autotransformers on supply systems that are solidly grounded constitute a violation of 215.11 and 450.5. And, this applies where the autotransformer is used to derive a neutral for neutral connected loads. While this is acceptable where the autotransformer is supplied from an ungrounded supply system, it is not acceptable where the supply system is solidly grounded. And, this applies where the autotransformer is supplied by the service or a separately-derived system.

The problem with this installation is that where the autotransformer is downstream of the grounding (earth) connection at the service or the separately-derived system, and a ground-fault occurs on this upstream system, the ground-fault will be shared by the upstream system and the grounded autotransformer. Serious damage to the autotransformer may be the result.

Impedance-grounded systems provide a means of reducing the effects of transient voltages that are associated with lightning and external power faults. This is certainly a benefit over an ungrounded system.

But, the fact still remains that in either an impedance-grounded system or an ungrounded system, a single ground-fault will not cause an overcurrent device to open. This may be desirable for certain types of installations, and sometimes this is an outright requirement. A ground detection system, as well as qualified maintenance, should ensure that a detected ground-fault is promptly located and repaired. If this is an arcing-fault (more often it will be) the ground-fault may soon become a short-circuit, where overcurrent devices should clear and there may be a possibility of equipment damage, to say nothing of a nonorderly shutdown of systems, which may also be a hazard.

If a phase-to-ground-to-phase fault occurs in the same enclosure (metal conduit, etc.) the impedance between these faults may be relatively low, and the overcurrent protection will very likely clear. However, if these opposite phase-faults occur in separate enclosures, the impedance factor will be markedly increased and current will flow between these two faults. This current flow may be high enough to cause significant equipment damage.

The impedance-grounded system has the advantage of having a ground (earth) reference point and this is certainly an advantage in limiting the voltage-rise on the system and equipment.

## Definitions

### *Effective Ground-Fault Current Path-

This term defines the ground-fault conducting path, from the point where the ground-fault occurs, to the source of the electrical supply (transformer, generator, etc.).

This effective ground-fault current path must have 3 components. These are:

1. The path must be permanent and continuous.
2. The path must have ample capacity to safely carry the ground-fault current likely to be imposed on it. This provision shows the importance of understanding the short-time ampere ratings of conductors, as the ground-fault current may be quite high, but typically of a short duration, until the overcurrent device operates to clear the ground-fault. The amount of ground-fault current that is expected to flow through this conducting path, and the duration of the ground-fault, or the time it takes for the overcurrent device to clear this fault, must be known in order to satisfy this requirement.

For example, let's say that the calculated ground-fault current on a 100 ampere circuit is 13,950 amperes. This circuit is protected by a 100 ampere, 3-pole molded-case circuit breaker, and the conductors are No. 3 THHN, copper (26.66mm²). The circuit breaker has a one-cycle (.016 seconds) (60Hz) clearing-time. The equipment grounding conductor for the 100 ampere circuit is No. 8 THHN copper (8.37mm²), taken from the minimum size equipment grounding conductor Table 250.122, based on the 100 ampere circuit rating. If we check the short-time ampere rating of this insulated No.8 copper wire from the Table in this Chapter, we find that the one-cycle insulation withstand rating is 6,912 amperes. However, the calculated fault-current in this circuit is 13,950 amperes. If a ground-fault occurred on this circuit, the insulation on this No. 8 AWG copper wire would be destroyed. This wire would not provide an effective ground-fault current path. Not only would this violate the second provision of this definition, but, also, this would be a violation of Section 110.10.

The solution would be to change the type of overcurrent device (faster clearing time), or increase the size of the equipment grounding conductor.

If we check the Table of conductor insulation withstand ratings, based on the one-cycle clearing time of the 100 ampere circuit breaker, we may select a No. 4 AWG copper wire, (21.15 mm$^2$), which has a one-cycle insulation withstand rating of 17,466 amperes, and this exceeds the calculated fault-current of 13,950 amperes in our example. This example shows why Table 250.122 only identifies the minimum sizes of equipment grounding conductors.

### Insulation Withstand Ratings 150° C. Maximum

| AWG | Normal 75°C. | 5 Second 150° C. | 1 Second 150° C. | 1 cycle – .016 sec. 150° C. | 1/2 cycle – .008 sec. 150°C. | 1/4 cycle – .004 sec. 150° C. | 1/8 cycle – .002 sec. 150° C. |
|---|---|---|---|---|---|---|---|
| 14 | 20A | 97A | 217A | 1,715A | 2,425A | 3,429A | 4,850A |
| 12 | 25A | 155A | 347A | 2,740A | 3,875A | 5,480A | 7,750A |
| 10 | 35A | 246A | 550A | 4,349A | 6,150 A | 8,697A | 12,300A |
| 8 | 50A | 397A | 888A | 6,912A | 9,775A | 13,824A | 19,550A |
| 6 | 65A | 621A | 1,389A | 10,978A | 15,525A | 21,956A | 31,050A |
| 4 | 85A | 988A | 2,209A | I7,466A | 24,700A | 34,931A | 49,400A |
| 3 | 100A | 1,245A | 2,784A | 22,008A | 31,125A | 44,017A | 63,450A |
| 2 | 115A | 1,571A | 3,513A | 27,772A | 39,275A | 55,543A | 78,550A |
| 1 | 130A | 1,981A | 4,430A | 35,019A | 49,525A | 70,039A | 99,050A |
| 1/0 | 150A | 2,499A | 5,588A | 44,176A | 62,475A | 88,353A | 124,950A |
| 2/0 | 175A | 3,150A | 7,044A | 55,685A | 78,750A | 111,369A | 157,500A |
| 3/0 | 200A | 3,972A | 8,882A | 70,216A | 99,300A | 140,431A | 198,600A |
| 4/0 | 230A | 5,008A | 11,198A | 88,530A | 125,200A | 177,060A | 250,400A |
| 250 Kcmi | 25SA | 5,917A | 13,231A | 104,599A | 147,925A | 209,198A | 295,850A |
| 300 Kcmi | 285A | 7,101A | 15,878A | 125,529A | 177,525A | 251,058A | 355,050A |
| 350 Kcmi | 310A | 8,284A | 18,524A | 146,442A | 207,100A | 292,884A | 414,200A |
| 400 Kcmi | 335A | 9,467A | 21,169A | I67,354A | 236,675A | 334,709A | 473,350A |
| 500 Kcmi | 380A | 11,834A | 26,462A | 209,198A | 295,850A | 418,395A | 591,700A |

### <u>Example</u>

### <u>Insulation Withstand Rating</u>

No. 6 AWG copper – 65 amperes (continuous) – 75°C
No. 6 AWG copper – 26,240 circular mils – 13.30mm$^2$
$I^2T$ – ampere – squared seconds
$I^2T$ – one ampere for every 42.25 circular mils of conductor cross- sectional area for 5 seconds

$$No.6\ AWG - \frac{26,240\ circular\ mils}{42.25} = 621\ amperes - 5\ seconds$$

To determine the insulation withstand rating for 1 cycle (.016 seconds), the calculation is as follows:

621 amperes × 621 amperes × 5 seconds = 1,928,205

$$\frac{1,928,205}{.016} = 120,512,812$$

$$\sqrt{120,512,812} = 10,978\,ampers$$

Therefore, the one-cycle insulation withstand rating is 10,978 amperes, which will produce a temperature of 150°C. This is the maximum temperature that the insulation can safely withstand without damage.

**Terminal Withstand Ratings 250° C. Maximum**

| AWG | Normal 75°C. | 5 Second 250° C. | 1 Second 250° C. | 1 cycle – .016 sec. 250° C. | 1/2 cycle – .008 sec. 250° C. | 1/4 cycle – .004 sec. 250° C | 1/8 cycle– .002 sec. 250° C. |
|---|---|---|---|---|---|---|---|
| 14 | 20A | 112A | 250A | 1,980A | 2,800A | 3,960A | 5,600A |
| 12 | 25A | 178A | 398A | 3,147A | 4,450A | 6,293A | 8,900A |
| 10 | 35A | 282A | 631A | 4,985A | 7,050A | 9,970A | 14,100A |
| S | 50A | 449A | 1,004A | 7,937A | 11,225A | 15,875A | 22,450A |
| 6 | 65A | 714A | 1,597A | 12,622A | 17,850A | 25,244A | 35,700A |
| 4 | 85A | 1,136A | 2,540A | 20,082A | 28,400A | 40,164A | 56,800A |
| 3 | 100A | 1,459A | 3,262A | 25,792A | 36,475A | 51,583A | 72,950A |
| 2 | 115A | 1,806A | 4,038A | 31,926A | 45,150A | 63,852A | 90,300A |
| 1 | 130A | 2,277A | 5,092A | 40,252A | 56,925A | 80,504A | 113,850A |
| 1/0 | 150A | 2,873A | 6,424A | 50,788A | 71,825A | 101,576A | 143,650A |
| 2/0 | 175A | 3,622A | 8,099A | 64,029A | 90,550A | 128,057A | 181,100A |
| 3/0 | 200A | 4,566A | 10,210A | 80,716A | 114,150A | 161,432A | 228,300A |
| 4/0 | 230A | 5,758A | 12,875A | 101,788A | 143,950A | 203,576A | 287,900A |
| 250Kcmil | 255A | 6,803A | 15,212A | 120,261A | 170,075A | 240,522A | 340,150A |
| 300 Kcmil | 285A | 8,163A | 18,253A | 144,303A | 204,075A | 288,606A | 408,150A |
| 350 Kcmi | 310A | 9,524A | 21,296A | 168,362A | 238,100A | 336,724A | 476,200A |
| 400 Kcmi | 335A | 10,884A | 24,337A | 192,404A | 272,100A | 384,808A | 544,200A |
| 500 Kcmi | 380A | 13,605A | 30,422A | 240,505A | 340,125A | 481,009A | 680,250A |

## Example

## Terminal Withstand Rating

No. 6 AWG copper – 65 amperes (continuous) – 75°C
No. 6 AWG copper – 26,240 circular mils – 13.30mm$^2$

$I^2T$ – one ampere for every 36.75 circular mils of conductor cross-sectional area for 5 seconds

$$No.6\ AWG - \frac{26,240\,circular\,mils}{36.75} = 714\,amperes - 5\,seconds$$

To determine the conductor terminal withstand rating for 1 cycle (.016 seconds), the calculation is as follows:

$$714\ amperes \times 714\ amperes \times 5\ seconds = 2,548,980$$

$$\frac{2,548,980}{.016} = 159,311,250$$

$$\sqrt{159,311,250} = 12,622\,amperes$$

Therefore, the one-cycle (.016 seconds) terminal withstand rating is 12,622 amperes, which will produce a temperature of 250°C. This is the maximum temperature that the terminal can safely withstand.

### Fusing Or Melting Current 1083° C. Maximum

| AWG | Normal 75° C. | 5 Second 1083° C. | 1 Second 1083° C. | 1 cycle – .016 Sec. 1083° C. | 1/2 cycle – .008 Sec. 1083° C. | 1/4 cycle – .004 Sec. 1083° C. | 1/8 cycle – .002 Sec. 1083° C. |
|---|---|---|---|---|---|---|---|
| 14 | 20A | 254A | 568A | 4,490A | 6,350A | 8,980A | 12,700A |
| 12 | 25A | 403A | 901A | 7,124A | 10,075A | 14,248A | 20,150A |
| 10 | 35A | 641A | 1,433A | 11,331A | 16,025A | 22,663A | 32,050A |
| 8 | 50A | 1,020A | 2,281A | 18,031A | 25,500A | 36,062A | 51,000A |
| 6 | 65A | 1,621A | 3,625A | 28,656A | 40,525A | 57,311A | 81,050A |
| 4 | 85A | 2,578A | 5,765A | 45,573A | 64,450A | 91,146A | 128,900A |
| 3 | 100A | 3,312A | 7,406A | 58,548A | 82,800A | 117,097A | 165,000A |
| 2 | 115A | 4,101 A | 9,170A | 72,461A | 102,475A | 144,922A | 204,950A |
| 1 | 130A | 5,169A | 11,558A | 91,376A | 129,225A | 183,105A | 258,950A |
| 1/0 | 150A | 6,523A | 14,586A | 115,311A | 163,075A | 230,623A | 326,150A |
| 2/0 | 175A | 8,221A | 18,383A | 145,328A | 205,525A | 290,656A | 411,050A |
| 3/0 | 200A | 10,364A | 23,175A | 183,211A | 259,100A | 366,423A | 518,200A |
| 4/0 | 230A | 13,070A | 29,225A | 231,047A | 326,750A | 462,094A | 653,500A |
| 250 Kcmil | 255A | 15,442A | 34,529A | 272,979A | 386,050A | 545,957A | 772,100A |
| 300 Kcmi | 285A | 18,530A | 41,434A | 327,567A | 463,250A | 655,134A | 925,500A |
| 350 Kcmi | 310A | 21,618A | 48,339A | 382,156A | 540,450A | 764,312A | 1,080,900A |
| 400 Kcmi | 335A | 24,707A | 55,247A | 436,762A | 617,675A | 873,524A | 1,235,350A |
| 500 Kcmil | 380A | 30,883A | 69,056A | 545,939A | 772,075A | 1,091,879A | 1,544,150A |

## Example

## Fusing or Melting Current

No. 6 AWG copper – 65 amperes (continuous) – 75°C
No. 6 AWG copper – 26,240 circular mils – 13.30mm²
I²T – one ampere for every 16.19 circular mils of conductor cross-sectional area for 5 seconds
I²T –ampere –squared seconds

$$No.6\ AWG - \frac{26,240\,circular\,mils}{16.19} = 1621\,amperes - 5\,seconds$$

To calculate the fusing or melting current for 1 cycle (.016 seconds), the calculation is as follows:

1621 amperes × 1621 amperes × 5 seconds = 13,138,205

$$\frac{13,138,205}{.016} = 821,137,812$$

$$\sqrt{821,137,812} = 28,656\,amperes$$

Therefore, the one-cycle fusing or melting current of this conductor is 28,656 amperes, which will produce a temperature of 1083°C. Based on these conditions, the conductor will fuse or melt in 1 cycle.

### AWG - To Metric Conversion Chart

| AWG | Circular Mil Area | Metric Size – MM² |
|---|---|---|
| 14 | 4110 cm | 2.08 |
| 12 | 6530 cm | 3.31 |
| 10 | 10,300 cm | 5.261 |
| 8 | 16,510 cm | 8.367 |
| 6 | 26,240 cm | 13.3 |
| 4 | 41,740 cm | 21.15 |
| 3 | 53,620 cm | 26.67 |
| 2 | 66,360 cm | 33.62 |
| 1 | 83,690 cm | 42.41 |
| 1/0 | 105,600 cm | 53.49 |
| 2/0 | 133,100 cm | 67.43 |
| 3/0 | 167,800 cm | 85.01 |
| 4/0 | 211,600 cm | 107.2 |
| 250 kcmil | 250,000 cm | 127 |
| 300 kcmil | 300,000 cm | 152 |
| 350 kcmil | 350,000 cm | 177 |
| 400 kcmil | 400,000 cm | 203 |
| 500 kcmil | 500,000 cm | 253 |

$$e.g. - No.2\,AWG = \frac{66,360\,Circular\,Mils}{1973.53} = 33.625mm^2$$

I would like to point out that if the fusing current of the No. 8 AWG copper wire was determined from the chart in this book on Page 11, the <u>fusing current</u> for one-cycle is referenced as 18,031 amperes. Once again, if our calculated ground-fault current is 13,950 amperes, then the minimum size equipment grounding conductor, No.8 AWG copper from Table 250.122, would be acceptable.

However, my suggestion is to use the chart based on insulation withstand ratings. This is because the equipment grounding conductor is within the same raceway or cable with the other circuit conductors (for important impedance reduction) (Section 300.3(B)), and if it were an insulated wire, this would serve to protect the other conductors from the thermal-stress of the ground-fault.

Another possible solution would be to use a current-limiting circuit breaker instead of the standard device. Current-limiting devices (fuses or circuit breakers) clear the fault in ½ cycle (0.008 seconds), or less. So instead of a 1 cycle (.016 seconds) clearing time for the standard circuit breaker, we would have reduced the fault-clearing time to no more than ½ cycle, and this would increase the insulation withstand rating for the No.8 AWG copper wire (8.37 mm$^2$) to at least 9,775 amperes. However, the available ground-fault current in our example is 13,950 amperes. So, the No.8 copper wire is not acceptable as an equipment grounding conductor for this circuit. Using a No. 6 AWG insulated copper wire (13.3 mm$^2$) for the equipment grounding conductor would be acceptable, as its ½ cycle insulation withstand rating is 15,525 amperes.

This is a good example of using the conductor short-time rating charts in this book to determine the validity of the effective ground-fault current path.

3. The ground-fault current path must be of sufficiently low impedance to limit the voltage-to-ground and facilitate the operation of the overcurrent device (on a solidly-grounded system). During a ground-fault, as the equipment grounding conductor is carrying the ground-fault current back to the source (transformer, generator, etc.), the voltage-drop associated with this current flow produces a voltage-rise, above earth potential (0 volts), on anything connected to the equipment grounding conductor. This may produce an excessive voltage-rise on equipment. Possibly a dangerous touch-potential may appear on equipment frames. So, the equipment grounding conductor must be properly sized to limit the voltage-drop and resultant voltage-rise. Where a change occurs in the size of the ungrounded conductors to compensate for voltage-drop, a proportional increase must be made in the size of the equipment grounding conductor (250.122(B)). This provision may not be applicable for DC circuits, unless they are excessively long. For example, 690.45

states that increasing the size of the EGC to compensate for voltage-drop for Solar Photovoltaic Source and Output circuits is not required.

## Voltage-Drop Example

Branch-Circuit Rating–40 Ampere
Circuit Voltage–240 volts–Single–Phase Circuit
Load 32 Amperes (Continuous)
Normal Ungrounded Conductor Size–No.8 AWG Copper
Minimum Equipment Grounding Conductor Size-No. 10 AWG Copper (Table 250.122)
Circuit length (from source to load)–270 feet–(82.3 meters).

According to Section 210.19 (A)-Informational Note No.4, voltage-drop for a combination feeder and branch-circuit should be limited to 5% of the applied voltage to provide for reasonable efficiency of operation. For a branch-circuit, the voltage-drop should be limited to 3%. For a feeder, this Informational Note expresses a voltage-drop of 3% of the applied voltage as the recommended limit for reasonable efficiency of operation.

This information is repeated in Section 215.2(A)(1), Informational Note No.2

Remember, as Informational Notes, this information is not mandatory (90.5(C)). So lower, or even higher levels of voltage-drop may be acceptable, or even necessary, depending on the type of equipment supplied. See 547.9(C) Informational Note for Agricultural Buildings, and 647.4(D) for Sensitive Electronic Equipment, where the voltage-drop for this equipment is limited to 1.5% for branch circuits and 2.5% for a combination feeder and branch circuit.

**Continuous Load**-210.19(A)(1),(a)-210.20(A)-A load where the maximum current is expected to continue for 3 hours or more (Article 100).

$$Voltage - Drop = \frac{2K \times L \times I}{CM\,(circular\,mil\,area\,of\,conductor)}(Single - Phase)$$

$$Voltage - Drop = \frac{1.732 \times L \times I}{CM\,(circular\,mil\,area\,of\,conductor)}(Three - Phase)$$

## Where

$$K = \frac{12.9\,ohms \quad copper\,wire}{21.2\,ohms - aluminum\,wire} \quad (ohms\,per\,circular\,mil\,foot,or\,ohms\,per\,mil\,foot)$$

This information is in accordance with the National Bureau of Standards Handbook 100 and Handbook 109.

L = one-way length in feet

I = amperes (Intensity) of load

(Section 220.5(A)-Unless other voltages are specified for the purposes of calculating branch-circuit and feeder loads, nominal system voltages of 120, 120/240, 208Y/120, 240, 347, 480Y/ 277, 480, 600Y/347, and 600 volts shall be used).

## Example

### No. 8 AWG

$$Voltage - drop = \frac{25.8 \times 270 \times 32}{16,510cm(No.8AWG)} = \frac{222,912}{16,510cm} = 13.5\,Volts$$

$Voltage - drop = 13.5\ Volts$

$$Circuit\,Voltage = \frac{240\,Volts}{7.2Volts} \times .03 \quad (3\% - 210.19(A) - Informational\ Note\ No.4)$$

### No. 6 AWG

$$Voltage - drop = \frac{25.8 \times 270 \times 32}{26,240cm(No.6AWG)} = \frac{222,912}{26,240cm} = 8.5\,Volts$$

$Voltage - drop = 8.5\ Volts$

### No. 4 AWG

$$Voltage - drop = \frac{25.8 \times 270 \times 32}{41,740cm(No.4AWG)} = \frac{222,912}{41,740cm} = 5.34\,Volts$$

$Voltage - drop = 5.34\ Volts$

A No.4 AWG-copper conductor-41,740 circular mils (21.15mm²) will limit the voltage-drop to less than 3% of the applied voltage (240 Volts × .03 = 7.2 Volts)

The equipment grounding conductor must be proportionately increased in size to compensate for voltage-drop (250.122(B)).

$$\frac{No.4\ AWG = 41,740\ circular\ mils}{No.8\ AWG = 16,510\ circular\ mils} = 2.53$$

40 ampere circuit-No.10 AWG copper equipment grounding conductor – Table 250.122

$$No.10 = \frac{10,380\,circular\,mils \times 2.53}{26,261\,circular\,mils}$$

26,261cm = No.4 AWG copper-41,740 circular mils

The equipment grounding conductor for the 40 ampere circuit is No.4 AWG copper.

A change in 250.122(B) for 2020 is a new Exception, where the increase in size of the EGC may be determined by a qualified person instead of a proportional increase in size. The AHJ will determine the qualifications of this qualified person.

To solve for the required circular mil area of the conductor to limit the voltage-drop to no more than 3%, based on the same circuit conditions, the equation is as follows (single-phase):

$$Circular\,mils\,area = \frac{2k \times L \times I}{7.2\,Volts\,(3\%\,of\,240\,volts)} = 30,960\,circular\,mils$$

* **Ground** – The earth (or to some conducting body that serves in place of the earth). As we will discuss in detail later, certain conducting bodies may serve as a ground reference. For example, the metal frame of a vehicle or the metal and the metallic skin and frame of an airplane.
* **Ground–Fault** – This is an accidental connection between an ungrounded conductor and the equipment grounding conductor, which may be a wire, metal raceway, metal cable tray, metallic equipment frame, or the earth (250.118).

Based on extensive research, it is safe to conclude that 90% of faults occurring in electrical systems are ground-faults. And, further, 90% of ground-faults are arcing-type faults. In addition, there is a voltage-drop associated with this arcing fault, which has an effect of reducing the amount of ground-fault current in the circuit. And couple this problem with the impedance of the ground-fault return path, and it is easy to see why the circuit overcurrent device may not operate in a short enough period of time, or, at all, to prevent a hazard to people, and/or, to equipment. And because the ungrounded conductors come in contact with the equipment grounding system in many places throughout a typical distribution system, it is easy to see why a ground-fault is the most common type of fault. It may not be caused by damage to the insulation, but simply, by contaminants that have been absorbed through the insulation, such as dirt combined with moisture.

* **Grounded** - This is a connection to ground or to a conducting body that connects to ground (earth).
* **Grounded, Solidly** – This is a physical connection to ground (earth) through a connection without a resistance element or impedance device in series with this connection (250.20(A),(B),(C)).

I used to ask a question in my grounding classes whether the equipment grounding conductor (metal raceway, metal cable tray, or wire) would carry any current during normal conditions. And the answer to this question was an emphatic 'no'. But, the actual answer is 'yes'. The equipment grounding conductor is installed within the same raceway, cable tray, cable, or cord with the other circuit conductors (AC Systems), or 690.43(C) for DC circuits for PV array circuit conductors. Or, it is the metallic raceway or cable tray itself (300.3(B), (300.5 (1)), (392.60(A). Therefore, some of the current flowing in the circuit conductors is capacitively-coupled into the equipment grounding conductor. This is known as a capacitive-charging current, and even on large systems, it should be quite low.

If higher current is present in the equipment grounding conductor, it may be indicative of upstream cross-bonds between a grounded (neutral) conductor and the equipment grounding conductor. Or, as I have seen in some installations, where the grounded neutral conductor is used as a means of grounding equipment enclosures. It should be noted that at least up to the 1993 NEC, the neutral conductor was permitted to be used to ground the equipment frames of electric ranges and clothes dryers (250.140). This was typically done through the use of Type SE cable, which has an uninsulated neutral conductor as part of the cable assembly. This practice was prohibited in the 1996 NEC, with an exception retained for existing installations (250.140, Exception), (250.142(B), Exception No.1). Also, for grounding meter enclosures on the load side of the service disconnect, the grounded circuit conductor (neutral) may be used for this purpose (250.142(B), Exception No.2).

* **Grounded Conductor** – This is a conductor that has been intentionally grounded. This conductor is, most often, referred to as the 'neutral' conductor. On single-phase, three-wire systems, and three-phase, four-wire, Wye connected systems, the system neutral point is solidly connected to ground (earth). The neutral point of these systems is the physical point, where voltage from the other points of the system would be equal. For example, a three-phase, Wye-connected transformer secondary, where voltage from X1-X2-X3 to the X0 (neutral point) would be equal. The X0 neutral point may be solidly grounded, and a grounded (neutral) conductor would originate from X0 (250.24, 250.26).

However, on a three-phase, Delta, corner-grounded system, one phase conductor is intentionally grounded. This conductor is a grounded conductor, but not a 'neutral' conductor. However, the means of identifying grounded conductors (200.6) applies to neutral conductors, as well as the grounded phase conductor of a corner-grounded Delta system.

Section 200.4 applies to the installation and identification of neutral (or grounded) conductors, where multiple circuits are installed in a common enclosure.

Section 200.6 and 200.7 specify the requirements for identifying grounded conductors, typically by the use of white or gray insulation or by white or gray markings at terminations.

## * Ground-Fault Circuit Interrupter-

This device was invented by Professor Charles Dalziel at the University of California, (Berkely), and he received a patent in 1961. A commercially available GFCI was introduced in 1964, specifically for wet-niche swimming pool lighting fixtures, where the leakage current could have been as high as 20ma. These GFCI's have not been listed for many years.

The first reference for GFCI's in the NEC appeared in 1968. This device has a differential transformer, which surrounds the supply and return conductors and monitors the current difference in these conductors. If the current difference reaches a level 4-6 ma or higher, (Class A GFCI), the unit trips in about 1/40 second. The listing standard for this device is UL 943.

The required use of these devices has expanded significantly over time. One of the original Code references, which addresses the use of GFCI's in dwelling and nondwelling occupancies, is Section 210.8.

## * Ground-Fault Current Path-

This is defined as a conductive path, from the point on the wiring system where the ground-fault develops, and then, through any conducting paths back to the source of the electrical supply. Hopefully, the lowest impedance path for this current to flow is through the equipment grounding conductor. However, other conducting paths may parallel the equipment grounding conductor. A portion of the ground-fault current may flow through these parallel paths. Also, when the ground-fault current reaches the service equipment (or ahead of the service equipment), the current will divide, unevenly, on its path back to the source. Due to the fact that there will be a grounding electrode connected to the grounded conductor and the equipment grounding system within the service equipment, or ahead of the service equipment, and, a grounding electrode connected to the neutral point of the supply transformer, another parallel path is established through the earth between these two grounding electrodes. Some of the normal neutral current will return to the transformer through this earth path, as well as some of the ground-fault current. This is a conducting path, albeit, a high impedance path, as compared to the path through the service grounded (neutral) conductor. It would be safe to say that only a small percentage of neutral current, or ground-fault current, will return to the source through the earth.

### * <u>Ground-Fault Protection of Equipment-</u>

This protection is a critical part of a solidly-grounded 3-phase, 4-wire, Wye-connected system, which operates at over 150 volts-to-ground, and not in excess of 1000 volts phase-to-phase. The most common type of system for this protection is a three-phase, 4-wire, solidly-grounded, Wye-connected electrical system with a voltage of 480 volts, phase-to-phase, and 277 volts, phase-to-neutral, and having a service disconnect rated at 1000 amperes or more. This is a type of distribution system where arcing ground-faults are the most destructive.

It is very important that the pick-up setting of the GFP is selectively coordinated with downstream overcurrent devices in order to prevent a nonorderly shutdown of an entire system. Sections 215.10 (Feeders), 230.95 (Services), and 517.17 (Health Care Facilities), address the requirements for this protection. Section 517.17 requires a second level of ground-fault protection on every feeder downstream of this protection at the service equipment. This downstream GFP will have a lower pickup setting than the main GFP to assure that the feeder device will open ground-faults on its load side, without affecting the operation of the service device. This will limit an outage to a single feeder, so that the rest of the distribution system will remain operational. In addition, Sections 700.6(D) and 701.6(D) require that a ground-fault <u>sensor</u> be provided for an emergency system or a legally-required standby system on solidly grounded Wye systems of more than 150 volts-to-ground and with overcurrent devices rated 1000 amperes, or more. This permits the ground-fault to be detected without interrupting the emergency system or the legally-required standby system and the fault cleared when the normal service is restored. For Critical Operations Power Systems (708.52(B)), an additional step of ground-fault protection at the next level of feeder disconnecting means downstream toward the load is required.

### * <u>Grounding Conductor, Equipment-</u>

This is the conductor, which may be in the form of a wire, metal raceway, metallic cable assembly, or metal cable tray. Section 250.118 identifies the various types of equipment grounding conductors. These conducting paths must be able to carry the maximum ground-fault current that may be imposed on them. If the equipment grounding conductor is in the form of a wire, Table 250.122 specifies its <u>minimum size</u>, in conjunction with Section 250.4 (A)(5) and 250.4(B)(4). Section 250.119 covers the identification of an insulated equipment grounding conductor. It is to have a continuous outer finish that is green, or green with one or more yellow stripes. However, where conductors are No. 4 AWG (21.15mm$^2$) or larger, other means of identification are acceptable, such as the use of green tape at terminations and other places where the wire is accessible, except in fittings, such as conduit bodies that contain no splices, or unused openings.

Sections 300.3(B) and 300.5(I) require that all of the conductors of the same circuit are to be installed together, and in close proximity to each other to maximize the effects of capacitive and inductive coupling and reduce the overall circuit impedance (AC Systems). Section 690.43(C) requires the equipment grounding conductors to be run with PV array circuit conductors where these conductors leave the vicinity of the PV array.

Magnetic flux density between conductors decreases with the square of the distance between conductors. So, it is important to keep the spacing between conductors to a minimum.

Check the UL guide on FHIT cable systems (e.g., cables that are installed on the life/safety and critical branches of an emergency system), where the equipment grounding conductors are installed in raceways and the equipment grounding conductor and raceway are a part of the electrical protection system.

### * Grounding Electrode-

This is the object that establishes a means of making a connection to the earth.

Section 250.52 identifies the various types of grounding electrodes that are used to make this earth connection.

It is very important that the grounding electrode, or grounding electrode system selected for a particular installation, be of a type that will assure a low resistance-to-ground. This concept is not only relative to the safety of people, but also, to assure the proper operation of equipment. This is especially true, due to the sensitivity of the sophisticated equipment in use today. And this may be more difficult now, as this equipment may be installed in an existing installation where a proper grounding system was not considered a priority.

In the past several years, I have had the privilege of working in many countries throughout Africa, as well as Haiti, and, even in several locations in North Korea. The installations were all health-care facilities. In the vast majority of these installations, a proper grounding system was not available, and, in some instances, there was no grounding system at all.

As an example, I was called upon to check the grounding system at the oldest government hospital in Addis Ababa, Ethiopia. GE Medical Systems was providing a new CT scanner for this hospital, and it has been my experience in other installations that GE requires a grounding electrode system to have a resistance-to-ground of less than 2 ohms. Depending on the type of soil, as well as other conditions, this low resistance-to-ground requirement can be difficult to achieve. I solved this problem at another hospital in Ethiopia, through the use of a ground ring, supplemented by several driven ground rods. But, that was a new installation and there were no area restrictions. This government hospital in Addis Ababa was in a confined area, so a ground ring was not an option. My preferred grounding electrode would have been a concrete-encased electrode. But this was an existing building, so there was no access to the rebar in the concrete footings (250.50, Exception).

Another option, that I had used at other installations, was a chemically-charged ground rod, which may have produced the desired low resistance-to-ground. But, this type of rod (UL 467J) was not available in Ethiopia.

And, to have one shipped from the U.S. would have taken at least six months, as the customs process in Ethiopia is, frustratingly, slow.

So, I decided to install a group of driven ground rods, 10 feet (3.048 meters) long x 5/8 inch (15.875 mm) in diameter. These were copper-clad steel rods, and they were properly spaced (a minimum of 2 rod lengths, or 20 feet (6.096 meters). (See the Informational Note following Section 250.53(A)(3)). In order to achieve a resistance-to-ground of less than 2 ohms, I had to install 13 ground rods. An engineer from GE performed a ground-resistance test to verify this low resistance-to-ground.

This may sound unrealistic, but before the installation of a driven or buried electrode, it is critical to perform a soil-resistivity test in order to determine the best location for the electrode or system.

There are three components that affect the resistance-to-ground of a driven or buried grounding electrode.

1. The metallic mass of the electrode. This portion of the resistance-to-ground is negligible. For example, the electrical resistance of a typical ground rod is quite low. A 5/8" diameter copper-clad steel ground rod has the approximate current-carrying capacity of a 3/0 AWG (85 mm$^2$) copper conductor (200 amperes at 75°C).
2. The resistance of the metal/soil interface. This is also quite low, unless there is a nonconductive coating on the metal mass of the electrode.
3. The resistivity of the soil surrounding the electrode. This is the component that will determine the resistance-to-ground of the grounding electrode. And, soil resistivity varies from place to place, and in some instances, even in the same area.

It would certainly be beneficial, and yes, very often critical, to locate an area where the soil resistivity is the lowest of the surrounding area, as that is the best location for the electrode. This value is expressed in ohm-centimeters. An ohm-centimeter is the electrical resistance across the faces of a cubic centimeter of soil. Where the soil resistivity is high, or unknown, soil enhancements are available. Bentonite (natural clay) has been used for this purpose. This material contains aluminum, iron, magnesium, and sodium. When this material comes in contact with water, it expands. Buried conductors may be surrounded with this material. Just like the typical soil surrounding a driven ground rod, the material within 6 inches (15.24 centimeters) of the rod or wire will have a dramatic effect on the total resistance-to-ground of the rod or conductor.

Erico 'GEM' (Ground-Enhancement Material) is even a better material, as its resistivity is lower than Bentonite (Erico.com) (1-800-248-weld). In addition, this material is noncorrosive, and it adheres to the rod or conductor more effectively than Bentonite. Its resistivity is about 10% of Bentonite.

Section 250.52(A) lists the acceptable grounding electrodes. They are:

1. A metal underground water pipe. In order to qualify as a suitable grounding electrode there must be at least 10 feet (3.048 meters) of metallic underground water pipe in direct contact with the earth. Since the 1978 NEC, underground metal water pipes must be supplemented by at least one additional grounding electrode. This is, very often, a ground rod. The reasoning here is that, even though the metal underground water pipe may be extensive, and, as such, the resistance-to-ground of this pipe may be quite low, if repairs to this pipe become necessary, some metallic portions of the pipe may be replaced with nonmetallic pipe.

   But, keep in mind that if a ground rod is used as the supplemental electrode, its resistance-to-ground must not exceed 25 ohms (250.53(A)(2)). If the resistance-to-ground exceeds 25 ohms, the single rod must be supplemented by an additional electrode, which may be an additional ground rod.

2. The metal frame of a building or structure which has been effectively grounded. This includes at least one vertical metal member of the building that is in direct contact with the earth for 10 feet (3.048 meters), or more. This structural metal member may, or may not, be encased in concrete. Or, the tie-down bolts which secure a structural metal member (steel column) to a concrete footing, with the tie-down bolts connected to the reinforcing rods in the footing.

3. A concrete-encased electrode. This grounding electrode is my personal favorite, as there is typically a very large amount of reinforcing rod, encased in concrete, in direct contact with the earth. As long as the concrete is not separated from the earth by a vapor barrier, or another type of nonconductive coating, this grounding electrode has a very low resistance-to-ground.

However, there are some limiting factors that must be considered in the use of this grounding electrode. These factors will be discussed later in this book.

4. Ground Ring. This is the grounding electrode that I used for the new Radiology building at Soddo Christian Hospital in Ethiopia. A new digital X-ray machine and a new CT scanner were to be installed, as well as 2-UPS units, and possibly an MRI sometime later. Once again, GE required a grounding electrode system with a resistance-to-ground of less than 2 ohms. The concrete footings were already poured, so I decided to use a ground ring. The ring consisted of a No. 2 AWG (33.63 mm$^2$) solid copper wire, buried to a depth of 3 feet (1 meter). The ground ring was supplemented by 8-5/8″ × 10′ ground rods. The length of the ground ring conductor was 300 feet (91.44 meters). A 3-phase, 380/220 volt, 165 kVA generator was installed at a later date, and the generator neutral point was bonded to this ground ring. So, everything in that building is at, virtually, the same potential so that there is no ground potential difference between the service supply and the generator.

5. Rod and Pipe Electrodes. Ground rods are very common, but also, they may be very ineffective. They are tested in accordance with UL 467. A copper-clad steel rod has an average service life of 30 years. This service life is dependent upon soil conditions, in that, if the soil has a low-resistivity, let's say, less than 10,000 ohm-centimeters, it is considered highly corrosive. From 10,000-30,000 ohm-centimeters, the soil is considered mildly corrosive. And above 30,000 ohm-centimeters, it is considered noncorrosive.

   The ground rod is not to be less than 8 feet in direct contact with the earth (2.44 m) and, typically 5/8″ (15.87 mm) in diameter, although 'listed' (UL 467) ground rods may be 1/2″ (12.7 mm) in diameter.

   A 3/4″ trade size (metric designator 21) galvanized pipe may be used, with the same minimum length of 8 feet (2.44 m) in direct contact with the earth. However, the service life of this grounding electrode may be limited, possibly, to no more than ten years.

6. Other Listed Electrodes are also permitted, such as a chemically-charged ground rod (UL 467J). These are hollow core rods and they are filled with various low-resistivity compounds, such as magnesium sulfate. The rod has an outside diameter of 2 1/8" (53.975 mm), and they are normally 10 feet (3.048 meters) in length. Although they can be made longer, and I have used them in an 'L' shape configuration. The compounds react with the normal moisture in the air, dissolve, and will slowly migrate into the soil through a series of holes in the shaft of the

rod, thereby adding minerals to the soil and lowering the soil resistivity. The performance of this rod will increase over time.

These rods may have a service life of up to 50 years. Where the soil resistivity has not been tested, 1 would strongly recommend that these electrodes be used.

7. Plate Electrodes. Plate electrodes, if of coated iron or galvanized steel are required to be at least 1/4" (6.4 mm) in thickness. If of nonferrous material, they may be 0.06" (1.5 mm) in thickness. They must expose at least 2 square feet (0.186 m$^2$) to exterior soil. These plate electrodes, while not commonly used, are typically buried on end as opposed to lying flat in order to reduce the size of the excavation.

250.52(B) addresses the materials and systems not permitted to be used as grounding electrodes.

1. A metal underground gas piping system. This is not necessarily due to an explosion hazard. It is because of the impressed current cathodic protection system that the gas supplier may be using to protect the gas pipe from galvanic corrosion. Connecting a grounding electrode conductor to the gas pipe could negate this protection due to the low level of AC current flowing through the grounding electrode conductor and into the gas pipe, and then through the earth on its path back to the supply transformer.

   Also, the gas pipe will typically have insulating joints. So, a long uninterrupted metal pipe may not be available.

2. Aluminum electrodes. Due to corrosion and oxidation, these electrodes would have a limited service life.

3. Structures and reinforcing steel that form the equipotential bonding for a swimming pool, the purpose of which is to reduce the voltage gradients (differences) in the pool area. This grid or structure is not a grounding electrode, and it may prove to be hazardous to persons in the pool area if it were a part of the grounding electrode system, due to inadvertent voltage variations that are associated with the grounding electrode system.

Section 250.53(A)(2) states that a single rod, pipe, or plate is to be supplemented by an additional grounding electrode, and the supplemental electrode may be any of the types referenced in Section 250.52(A)(2) through (A)(8). There is an exception here that waives this requirement if the single, rod, pipe, or plate has a resistance-to-ground (earth) of 25 ohms, or less. I have always maintained that this 25 ohm value is too high to provide proper protection for people, systems, and equipment. Especially, if there was a problem with lightning. If the service supply was affected by an indirect lightning strike, or, more likely, at an outside transformer supplying this distribution system, and the lightning current, of say, 20,000 amperes were carried into this ground rod with a 25 ohm-to-ground resistance, the voltage-rise, above earth potential, would be 500,000 volts. While this would be an instantaneous event (lightning currents reach a peak in about 2-10 microseconds), this, certainly, would be hazardous to people, and damaging to equipment. It should be noted that the 25 ohm resistance-to-ground of the ground rod would limit the level of fault current that may be carried into the earth. But, in order to effectively dissipate this current more effectively, and reduce the voltage-rise above earth potential on systems and equipment, a lower resistance-to-ground electrode is important.

It is extremely important that where multiple grounding electrodes are installed that they be bonded together to form one common grounding electrode system (250.50), (250.58). This includes the grounding electrodes that are installed for the strike termination devices of a lightning protection system (250.60, 250.106). The bonding of all separate grounding electrodes will serve to limit potential differences between them, and the systems and equipment which are connected to ground through these grounding electrodes. Also, the bonding of electrodes of different systems is addressed in 800.100(D), 810.21(J), 820.100(D), and 830.100(D).

Another benefit associated with the bonding of the grounding electrodes of different systems is that the overall resistance-to-ground of all of the connected systems may be markedly reduced, as these systems will now be in parallel. This will provide added protection for all of these systems.

## * Grounding Electrode Conductor-

The grounded conductor and the equipment grounding conductor are connected to the grounding electrode (system) by the grounding electrode conductor. Section 250.62 states that this conductor may be copper, aluminum, or copper-clad aluminum. And the conductor may be solid or stranded, insulated, covered, or bare. However, 250.64(A) prohibits aluminum or copper-clad aluminum conductors in direct contact with masonry or the earth, or where subject to corrosive conditions. And, where used outside, these conductor types may not be terminated within 18" (450 mm) of the earth.

Generally, this conductor must be installed in one continuous length. However, exothermically welded connections or compression (crimped) type connections are acceptable, as these connections are irreversible, which means that they cannot be taken apart with tools. Busbar connections are also acceptable (250.64(C)).

Where multiple service disconnects are installed as a group in separate enclosures (230.71), taps from each disconnect to the main grounding electrode conductor are permitted. This would be common where these separate enclosures are connected to an auxiliary gutter (Article 366), with the splices inside this gutter (250.64(D)).

Enclosures for grounding electrode conductors should be nonmetallic. However, steel enclosures are acceptable for physical protection if this enclosure is bonded at both ends to the internal grounding electrode conductor. This may be accomplished through the use of bonding bushings on each end of the steel enclosure. The size of the bonding jumper would be the same size as the internal grounding electrode conductor. The reason for this requirement is due to the magnetic properties of the steel enclosure. If significant current is flowing through the grounding electrode conductor, the magnetic field surrounding this conductor would induce a current in the steel enclosure which would be in a direction opposite the current flow through the grounding electrode conductor. This condition would produce an 'inductive choke' on the grounding electrode conductor which would significantly increase the impedance of the earth connection and possibly destroy the grounding electrode conductor in the process (250.64(E)).

Where the steel enclosure is bonded to the internal conductor on both ends, the effects of this 'inductive choke' are greatly reduced, but not completely eliminated.

The steel enclosure (conduit) may be connected by means of grounding-type locknuts or bonding-type bushings with properly sized bonding jumpers (250.92(B)).

If nonferrous enclosures are used for the physical protection of the grounding electrode conductor, bonding the conductor on each end of the enclosure is not necessary, due to the nonmagnetic properties of this material (250.64(E)(1)).

The sizes of AC Grounding Electrode Conductors are listed in Table 250.66 (for Direct Current Systems, Table 250.166 applies). This size is based on the size of the ungrounded service-entrance conductors, or the size of the ungrounded feeder conductors that extend from the source of a separately-derived system. From data that I studied 40 years ago, I have maintained that the wire sizes expressed in Table 250.66 are based on lengths up to 100 feet (30.48 meters). And a specific voltage-drop that is associated with this length. Because this voltage-drop is directly proportional to the voltage-rise on connected systems and equipment during fault conditions, the grounding electrode conductor should be increased in size for lengths of over 100 feet (30.48 meters).

A No. 6 AWG copper (13.3 mm$^2$) or No. 4 AWG aluminum (21.66 mm$^2$) conductor may be used as the sole connection to a ground rod, pipe, or plate electrode (250.66(A)). However, the use of an aluminum conductor for this purpose is limited due to the fact that aluminum is not to be terminated within 18 inches (45.72cm) of the earth (250.64(A)).

For concrete-encased electrodes, the connection to the reinforcing steel within a concrete footing may be No. 4 AWG copper (21.66 mm$^2$), (250.66(B)). However, due to galvanic action, be sure that the connecting device is listed and identified for this connection, that is, copper to steel, or an exothermically welded connection may be used.

# 2

# A Practical Example

As we begin our analysis of grounding electrical systems and equipment, 1 would like to place the emphasis on the importance of a complete understanding of each of the topics which comprise our study of this complex subject.

It seems to me that the misconceptions of the purposes of grounding systems and equipment have needlessly perpetuated unsafe electrical installations with respect to the protection of people, systems, and equipment.

Indeed, I have witnessed electrical installations in many health care facilities at various locations throughout Africa, Haiti, and even, North Korea that were woefully inadequate, and, even downright dangerous. These conditions are typically caused by inadequate training of personnel, and the lack of knowledge of available equipment and materials. Sadly, it can be extremely difficult, and sometimes, virtually impossible to correct these problems. Recently, I was called upon to install a new 200 kVA Voltage Regulator at a hospital in Ethiopia. The supply transformer was pole-mounted, and a set of supply conductors (3 Phase, 4 Wire, 380/220 Volt, Wye) extended underground to the service equipment. There was a junction box in this supply system which contained a set of fuses, and, what appeared to be a connection to a grounding electrode (of some sort). For some reason, the utility does not connect the neutral bushing of their supply transformers to a grounding electrode, but they prefer a downstream connection, which then becomes the responsibility of the customer. Is there any wonder why this vital earth connection is not made? And, if it is, it may consist of a conductor that is buried a few inches below grade. In checking the resistance-to-ground of the existing grounding electrode, I determined it to be 39 ohms. And this installation is in an area which is subject to frequent thunderstorm activity!!! At the service equipment location, consisting of a panelboard, with a main circuit breaker (service disconnect), there was a three-phase surge-protective device, with one-phase completely inoperative. The hospital generator (150 kVA/120 kW), equipped with a 4-pole transfer switch (a separately-derived system), had no neutral connection to ground. It is certainly imperative to

29

reference the generator neutral to the same grounding electrode used for the service equipment, and not to another grounding electrode in order to achieve the same ground (earth) potential for both systems.

So, to correct these serious errors, a chemically-charged ground rod was installed, equidistant from the neutral connection in the new voltage regulator cabinet (which also contained the new service disconnect, consisting of a 400 ampere, 3-pole circuit breaker), and the generator distribution cabinet, where the generator neutral connection was accessed.

In addition, two 10-foot (3 meter) long, copper-clad steel ground rods were installed in parallel and separated by twice their driven depth (20 feet, or 6 meters) for maximum efficiency (NEC 250.53(A)(3), Informational Note). These paralleled rods were installed and connected to the supply neutral in the junction box on the secondary side of the supply transformer that we alluded to earlier.

Now, with these ground rods connected in parallel with the chemically-charged ground rod for the service equipment and generator, the measured resistance-to-ground was reduced from the original 39 ohms to a value of 2 ohms. And, this low resistance-to-ground will be further reduced over time as the 'salt' (typically magnesium sulfate) from the chemically-charged ground rod migrates through the soil and lowers soil resistivity in the process.

I would also like to mention that the existing surge-protective device was replaced with a reactance-type surge-protective unit which was supplied and connected by the manufacturer of the Voltage Regulator. This SPD (Type 4) was installed inside the Voltage Regular cabinet, in close proximity to the incoming supply conductor terminals. So, the connecting leads are very short, and virtually straight. This, as compared to the original surge-protective device, where the connecting leads were at least 25 feet (8 meters) long, with several unnecessary bends in these wires, rendering them totally ineffective. Suffice it to say that for each one foot (1/3 meter) of unnecessary length in the surge-protective device conductor connections, 1000 volts of protection is compromised. In this installation, the original surge-protective device could have been connected with conductors of less than 3 feet (1 meter). But due to their 25 foot length, (8 meters), at least 22,000 volts of protection was lost (NEC 285.12).

Keep in mind that one of the purposes of providing a proper grounding system is the protection from lightning induced influences (which at this hospital location is almost a constant threat), as well as transient conditions, which may produce momentary variations of voltage, current, and frequency caused by utility faults associated with switching and capacitor functions.

In addition, due to the inconsistency of the incoming power system, where the utility supply is interrupted and restored as many as ten times per day, transients are a constant threat. And a rapid succession of transients on the normal system voltage wave produced by these external causes, as well as internally by motors, motor control relays, arc welders, and other control relays are classified as 'electrical noise' or electromagnetic interference (EMI).

In many cases, electrical noise may be controlled by the shielding of cable assemblies. But the interference produced by these causes, where the frequency of the electrical noise (EMI) is less than 3000 Hertz, cannot be cancelled by shielding. One possible solution is through the use of steel raceways (conduits) where 95% of the electromagnetic interference is eliminated.

Due to these excruciating circumstances, it is easy to see the need for a grounding system which has a negligible resistance-to-ground, as this is the first line of defense against these conditions.

I would like to emphasize that the method of making the connection to the grounding electrode (system) is extremely important in limiting the voltages associated with lightning and external power faults. Limiting the length of the grounding electrode conductor(s), as well as avoiding unnecessary, and especially sharp bends in this conductor(s), will serve to reduce the effects of imposed voltages (NEC 250.4(A)(1), Informational Note No.1).

The second line of defense against erratic and imposed voltages and associated surge currents at this hospital is the installation of the 200 kVA Voltage Regulator and the factory installed surge-protective device. This device will serve to attenuate the voltage and associated surge current. The location of this surge-protective device at the incoming service equipment location is classified as Location Category C (ANSI/IEEE Std. 62.41).

If further protection is necessary downstream of this location, such as at a local area panelboard, this SPD protection is considered Location Category B.

And, if it is determined that protection is needed at the connected equipment, this is considered Location Category A.

NEC Sections 285.21-23-24-25 identify Types 1-2-3 and 4 surge-protective devices.

Subsequently, while replacing the surgical lighting units in 4 operating rooms of this hospital, I noticed a voltage difference, and resultant current flow in the equipment grounding conductor of the supply circuit. I measured a current flow of 4 amperes on this equipment grounding conductor. So the entire grounding system in this facility is suspect and appropriate repairs must be made.

This storyline is quite common in these hospital installations throughout Africa (and elsewhere). But, I dare say that there are installations in the U.S.A. where certain inadequacies, either due to improper design and/or installation practices, have led to facilities that are unsafe and unreliable.

This introduction is meant to prepare the reader for a comprehensive summary of the 'History and Mystery' of grounding equipment and systems. This approach is to offer insight into the ways and means to accomplish the goal of providing a safe system, and one that assures the reliability and continuity of service that is essential in today's electrical environment.

# 3

# The Lightning Phenomenon

In 1754, Benjamin Franklin used his kite and key experiment to prove the existence of electrical charges in thunderclouds. He was quite fortunate to survive this experiment, as other people who attempted to do the same experiment did not survive.

Franklin developed a concept of lightning protection through the use of his 'lightning rod', or, as we know it today, the 'air terminal' or strike termination device. He argued that this lightning rod should be 'pointed' as opposed to being 'blunt'.

At the time, the concept of lightning protection was directly associated with the prevention of fire. Buildings, especially those that were tall, or had portions that were relatively high, such as church steeples, were made of wood.

Surely, the prevention of fire is still an important consideration. But, other concerns must also be considered. Such as the protection of people and equipment, preserving critical data, and limiting downtime.

Certainly, a lightning protection system for a building or structure is more important in areas that have a great deal of thunderstorm activity.

Isokeraunic maps have been developed by the National Weather Services of the National Atmospheric Administration. These maps identify the thunderstorm activity in different areas of the United States and Canada. Numbers are referenced on the map that are associated with what is known as 'thunderstorm days'. A thunderstorm day is one where thunder is heard at least once during the day.

The isokeraunic map of the United States shows that the most activity is in Central Florida (90 thunderstorm days), and the least activity in the contiguous United States is in Southern California (5 thunderstorm days).

This data may serve as an aid in determining the need for a lightning protection system, and, where this type of protection may not be of serious concern.

As a bit of trivia, the term 'isokeraunic' is taken from two Greek words, 'iso', meaning 'same', and 'keraunis', meaning 'thunderbolt'.

More accurate data relating to lightning strike conditions, typically over a 5 year period, is available.

There are differing schools of thought about the propagation of lightning, and how a typical lightning protection system actually works.

Let's examine this phenomenon and develop a better understanding of these causes and effects.

First, lightning is nature's greatest manifestation of 'static electricity'. And, as in other examples of static electricity, it is predominately direct current. However, there is an alternating current present, as well. And this alternating current has a frequency that ranges from 3 kHz to 10 mHz. It is this alternating current component, and its relatively high frequency, that establishes the reason for limiting the length of the conductors associated with Surge-Protective Devices and Surge-Arresters, as well as avoiding any unnecessary and especially sharp bends in these conductors (NEC 280.12 and 285.12). This also applies to the installation of grounding electrode conductors that connect to the grounding electrode system. By limiting the length and the number of bends, (especially sharp bends) in these conductors, the impedance is markedly reduced, and the voltage-rise and associated surge-current is restricted (NEC 250.4 (A)(1) Informational Note No.1).

Now, we know that electrical charges accumulate in cloud masses, typically due to the rapid contact and separation of different materials (dry ice, snow, dirt, chemicals, various contaminants, etc.). These electrical charges increase dramatically during violently turbulent conditions, that is, the level of static electricity increases with the speed of the contact and separation of these materials. By far, the vast majority of thunderclouds are negatively charged (more than 90%). These thunderclouds may be charged to as many as tens, or even hundreds of millions of volts. But, in a positively charged cloud, that is, where the charges are in the uppermost reaches of the cloud, the lightning strikes that are produced usually have much higher currents that those from negatively charged clouds. While, these currents are of extremely high magnitude, but of a short duration (2–10 microseconds), any lightning protection system must be designed to handle these high currents, taking into consideration the increased inductive reactance of the conducting paths due to the higher frequency of the alternating current component of lightning (3kHz–10mHz). As this charged mass moves overhead, opposite charges are induced on the earth. The only thing that separates the cloud borne charges from those on the earth is air, and its dielectric properties. As the charges continue to accumulate, the magnetic field surrounding the cloud becomes strong enough to cause the air to ionize. In so doing, its dielectric properties weaken, and the cloud borne charges begin to flow through the ionized air, typically in a zigzag pattern, with each section having a length of 100–150 feet (30.48–45.72 meters), and a current density of approximately 100 amperes. This conducting path is known as a 'stepped leader'. As it continues to form and it gets closer to the opposite earthbound charges, a point is reached where these earthbound charges flow upward to meet the stepped leader. A bright flash appears as the circuit is completed. There may be additional lightning flashes, until the cloud and earth charges are neutralized.

The lightning flash (lightning channel) heats the surrounding air and the rapidly expanding heated air creates an enormous pressure wave with the force of an explosion (thunder). This force, in itself, may cause significant damage to nearby structures.

Lightning may occur within clouds (anyone who has flown in an airplane through a storm has witnessed this activity), or from cloud to cloud, or from a cloud to the earth.

In our description of the formation of the stepped leader(s) and return stroke(s), we can examine theories that are associated with the protection of buildings and other structures.

NFPA 780 (Standard for the Installation of Lightning Protection) identifies the means of protection in great detail. Since Benjamin Franklin designed the first 'air terminal' as a protection method, there have been several variations in its design. He stressed that the air terminal should be pointed. Whereas, in modern theories, a blunt air terminal appears to be more effective.

The Franklin theory is centered around a passive means of protection, whereby, the air terminals tend to dissipate, or 'bleed off' the earthbound charges at the points of the air terminals, thereby removing the opposite charges on the building or structure and making it much less likely to be a target for the return stroke. I might add that this has been my belief for many years. During this activity, the 'bleeding off' of these charges at the points of the air terminals may even be visible.

Another theory is that the air terminals attract the stepped leader stroke. And these bonded air terminals and their associated 'lighting down conductors', as well as the grounding electrode system provide a preferred path to ground, as opposed to other conducting paths. This is a more active concept than the 'bleeding off' theory mentioned previously.

There is a 'dissipation array' lightning protection system that appears to work quite well, especially in areas with a great deal of thunderstorm activity. In this case, possibly hundreds of air terminals are exposed to provide protection for buildings, as well as other structures.

Yet another theory considers the formation of a corona cloud above the air terminals, which makes them less attractive to the stepped leader as it seeks to find a path to the earth.

In any event, none of these protection systems affords complete protection against lightning. But, regardless of the theory involved, lightning protection should be an important consideration in those regions of greater thunderstorm activity.

NFPA 780 specifies the location of strike termination devices, for example, every 20 feet (6.1 meters) where the strike termination devices have a length of 10 inches (25.4 centimeters), or a spacing of 25 feet (7.62 meters) where the air terminals have a length of 24 inches (60.96 centimeters). These strike termination devices are effectively bonded, such as around the periphery of the top of the structure (and additional points for larger structures) and then connected to the earth through a series of lightning down conductors, which may be spaced up to 100 feet (30.48 meters) apart. A closer spacing, possibly 50 feet (15.24 meters), will, very likely, provide better protection.

The heart of the lightning protection system is the grounding electrode, which establishes the '0' volts reference (earth) point.

In my opinion, the most effective grounding electrode for this protection system is a 'ground ring'. This grounding electrode consists of a No. 2 AWG (33.63 mm$^2$), bare copper conductor, buried to a minimum depth of 2½ feet (.762 meters), which completely encircles the building or structure. The minimum length of this conductor is 20 feet (6.0 meters) (NEC 250.52 (A)(4)). The decision to supplement the ground ring conductor with driven ground rods to lower its resistance-to-ground, as well as using a solid, <u>tinned</u> conductor, for longer service life, and using a larger conductor than identified in NEC 250.52 (A)(4), is strictly up to the designer (installer). Remember, the NEC is not intended as a design specification (NEC 90.1(A)).

The ground ring serves a purpose of equalizing the voltage gradient around the building or structure. This is certainly desirable, whether it is installed for lightning protection, or for the protection of people and equipment.

A concrete-encased electrode is also quite effective, especially where the structural steel columns are effectively bonded to the reinforcing rods in the concrete footing. And these may be used as a 'lightning down conductor' for the structure (NEC 250.52 (A)(3)). A note of caution however must be considered in the use of this electrode. It has been determined that as the current reaches from 500 to 2600 amperes, complete destruction of the concrete is possible. The concrete in a typical footing is hygroscopic, in that it tends to absorb and retain moisture from the soil surrounding it, even in virtually dry soil. In fact, this surrounding soil becomes a part of the electrode. Concrete has a resistivity which ranges from 3000 to 9000 ohm-centimeters. This is quite low, as compared to the soil surrounding the footing. A problem may develop if a relatively high fault current passes through the concrete footing, and the resultant heat energy causes the rapid expansion of the absorbed moisture in the footing as it changes to steam. Again, total destruction of the concrete footing may be the result. This problem has been an issue for electric utilities for many years. Significant damage to the concrete footings of transmission towers have been reported, either caused by the lack of continuity between the anchor bolts and the reinforcing rods of the footing, or the lack of an external connection to an auxiliary grounding electrode (ground rod).

For many years, the protection technique for concrete-encased electrodes has been to connect a copper conductor (No. 4 AWG ($21.15 \text{ mm}^2$), or larger, from the reinforcing rods in the footing to an external grounding electrode, which may be one or more ground rods. Connections may be made around the periphery of the footing or foundation. The connections to the external grounding electrode(s) will provide another path for the ground-seeking current to flow into the earth, instead of flowing solely through the absorbed moisture in the concrete.

The supplemental grounding electrode(s) for the concrete-encased grounding electrode will also serve to reduce the resistance-to-ground of the grounding electrode system, which is certainly beneficial.

One means of accessing the rebar in the concrete footing or foundation is to extend one or more reinforcing rods through the concrete (above grade) for connection to the grounding electrode conductor. However, the connection of the copper grounding electrode conductor to the exposed steel rebar may be a violation of 110.14, due to the intermixing of dissimilar metals, unless the connecting device is listed and identified for this purpose and, as we have mentioned in Chapter 1, an exothermic welded connection is acceptable.

In our summary of the major electrical problems with the electrical distribution system at the hospital in Ethiopia, I emphasized the need to provide protection in the form of Surge-Protective Devices. This equipment is critical in providing protection against transients from outside areas of disturbance.

Generally, Surge-Arresters are a part of the utility supply system. In areas of significant thunderstorm activity, or where topography may be a factor in deciding on the need for additional protection in the form of surge-arresters on supply systems of over 1000 volts (NEC Article 280), we have to recognize that it is not uncommon to see voltages of several thousand volts on 120 volt premises wiring systems. We recognize that exposed utility lines act as an antenna for induced voltages and surge currents associated with direct, or even nearby (indirect) lightning strikes. In addition, utility switching operations often lead to significant overvoltages. The standard for Surge-Arresters is ANSI/IEEE C62.11. A review of this standard, as well as NEC Article 280 is highly recommended.

As the transient surge passes through the arrester, a wave will flow in both directions, that is, into the distribution system and into the building supply system (transformer). Therefore, the surge-arrester must be installed in close proximity to the transformer and referenced to ground through conductors that are as short and straight as possible (280.14). This will significantly reduce the transient rise. This transient wave will increase in magnitude as the distance from the arrester also increases.

Some important aspects of providing protection through the use of Surge-Protective Devices should be noted here. Keep in mind that these devices, while carefully designed and listed (UL 1449), can, and do fail, usually in a short-circuit mode. Providing overcurrent protection for these devices, as some manufacturers suggest, may be a good idea, especially where the equipment is frequently monitored and the source of a failed component may be promptly corrected.

In addition, when a voltage variation causes the operation of the SPD, a large surge current will be directed into the grounding conductor to which the SPD is connected. There will be a momentary voltage-rise on this grounding conductor, which may be significant, albeit of very short duration. Any other equipment that is connected to this grounding conductor will also be affected by this condition. The best method of limiting this momentary voltage-rise is to provide a grounding electrode system that has a low resistance-to-ground.

The voltage transient or variation that the SPD is designed to 'clip' should be set at a value that is above the peak line-voltage, or 1.414 of the effective (rms) value of the voltage. For example, for a nominal voltage of 120 volts, the SPD should have a protective level that is <u>above </u>170 volts (120 × 1.414=169.68 volts).

The upper limit of the effective voltage level (rms) that the SPD can safely withstand will be identified as the Maximum Continuous Operating Voltage (MCOV).

An addition to the 2014 NEC requires that a listed surge-protective device be <u>in</u> or <u>on</u> all emergency system switchboards and panelboards (Section 700.8).

# 4

# The Grounding Electrode System

As we begin our study of grounding electrodes, we need to stress some very important topics. First, let us examine the definition of the term **Ground** from NEC Article 100. This term is defined as, the earth. At one time, this term was defined as, 'the earth, or some conducting body that serves in place of the earth.' I prefer the older definition for a very important reason. Yes, the earth is a reference point for the connection of electrical systems and equipment. The purpose of making this connection to earth is normally twofold. That is, to stabilize system voltages to ground during normal or abnormal conditions, and to limit the voltage-to-ground on electrical systems and equipment (250.4(A),(1)).

However, a direct connection to the earth is not always necessary to establish a ground or earth reference point for systems and equipment. For example, aircraft have no physical earth connection. And yet, they have grounded circuits. Equipment is also protected from touch potential differences without an earth connection. The aircraft frame and outer skin serve as the ground reference point. And everything is so well bonded that potential differences are, virtually, nonexistent. Even lightning currents are allowed to pass through the structure without affecting the integrity of the aircraft, or even sensitive circuits that are critical to its operation.

Also, the frame of a vehicle, car, truck, bus, etc., serves as a ground reference point in the same manner as the aircraft frame, in lieu of a physical earth connection.

And, NEC 250.34(A) and (B) identifies a means of using the frame of a portable or vehicle-mounted generator as a ground reference point, without a physical connection to a grounding electrode (earth).

In these cases, the generator supplies cord-and-plug equipment from receptacle outlets that are mounted on the generator. And the equipment grounding terminals of the receptacle outlets are bonded to the generator frame. Or, for equipment which is mounted on the generator, the frame of this equipment is bonded to the generator frame.

For generators mounted on Vehicles, an additional requirement specifies that the generator frame must be bonded to the vehicle frame (250.34(A),(B)).

39

The reason for the use of the portable generator frame or the vehicle frame as the ground reference in lieu of a physical earth connection is related to the <u>safety</u> of a person using cord-and-plug connected equipment which is supplied through receptacles on the generator, or equipment which is mounted on the generator or vehicle. Consider a portable generator that has its frame connected to a grounding electrode (ground rod), and a person using a portable tool connected into a receptacle on the generator. If there was a ground-fault in the portable tool, there would be 2 paths for the ground-fault current to flow. One path would be through the equipment grounding conductor in the supply cord, and the other path would be through the person and through the earth to the ground rod, and then through the grounding electrode conductor to the generator frame, and, finally, to the generator winding. This is certainly a serious safety issue, and the reason that 250.34(A),(B) permits the generator or vehicle frame to serve as the ground reference. OSHA Grounding Requirements for portable and vehicle-mounted generators also recognizes that the frames of this equipment need not be grounded (connected to the earth), and the frame may serve as the ground reference point (OSHA 29 CFR 1926.404(f),(3),(i)), 29 CFR 1926.404(f),(3),(i)(A), 29 CFR 1926.404 (f),(3),(i),(B).

Of course, this problem would be far worse if the supply cord did not have an equipment grounding conductor.

Section 590.6(A)(3) (temporary installations) requires that all 125-volt and 125/250 volt, single-phase, 15-20- and 30 ampere receptacle outlets that are a part of 15kW or smaller portable generators must have listed ground-fault circuit interrupter protection for personnel. This provision is intended to assure protection for personnel during construction or maintenance activities.

## And now, let's examine different types of grounding electrodes from NEC 250.52(A)

1. **A Metal Underground Water Pipe,** that is in direct contact with the earth for 10 feet (3.0 meters) or more, which may include any well casing which is bonded to the pipe. This grounding electrode has been used for decades. And, it stands to reason that an extensive metal underground water pipe would have a resistance-to-ground that is very low. Of course, where this type, or any type of grounding electrode is used at the service equipment, a small percentage of the current flowing through the neutral conductors at the service equipment, will flow into the grounding electrode conductor and into the water pipe, and then through the earth to the grounding electrode at the utility supply

transformer. And, eventually, to the transformer neutral point. This path is in parallel to that of the service grounded (neutral) conductor, and it is unavoidable. This conducting path is of high impedance, as compared to the service grounded (neutral) conductor path. And, even though there may be a significant number of connections to the water pipe, this does not create any hazard to people, or to the water piping system.

Of course, the practice of using nonmetallic water pipe has been with us for many years. The 1978 NEC was modified to require that the metal underground water pipe be supplemented with at least one additional grounding electrode. In many cases, this additional grounding electrode may be in the form of a ground rod, if the single rod has a resistance-to-ground of 25 ohms or less. Otherwise, an additional grounding electrode must be used. Of course, any other grounding electrode may be used for this purpose (NEC 250.53 (D)(2)).

Some local authorities require that where ground rods are used as the supplemental grounding electrode, there must be at least two rods in parallel, and they must be no more than six-feet apart (2.0 meters). I have no problem with parallel ground rods. In fact, it may be necessary, at times. My problem is the six-foot spacing between the rods. The paralleling efficiency of ground rods is increased by spacing them at twice the length of the longest rod. (NEC 250.53 (A)(3), Informational Note). For example, two, ten-foot (3.0 meter) ground rods should be separated by twice their length, or twenty feet (6.0 meters). At this spacing, their combined resistance-to-ground is approximately 50% of the single rod. Closer spacing will increase this resistance-to-ground.

2. **The Metal Frame of the Building or Structure,** which has been connected to the earth by extending at least one vertical structural metal member into the earth for 10 feet (3.0 meters) or more, with or without concrete-encasement. Or, by securing the structural steel tie down bolts to a concrete-encased electrode. This may be accomplished by welding the tie down bolts to the reinforcing rods in the concrete footing, or by the exothermic welding process, or the typical tie-wire method that is used for the reinforcing rod connections. It is very important that the structural metal frame of a building or structure be effectively connected to the earth. Where the outer building columns are properly bonded they may serve as lightning down conductors. In addition, they may serve as the earth connection for separately-derived systems (transformers generators, etc.) within the building or structure (250.30(A)(4)). And, the proper bonding and grounding of the structural steel will reduce the effects of touch potential differences.

3. **A Concrete-encased Electrode,** consisting of at least twenty feet (6.0 meters) of one or more bars of zinc, or other electrically conductive coated steel reinforcing bars or rods of not less than one-half inch (13 mm) in diameter (No. 4 rebar), installed in one continuous length. Or, in multiple pieces that are connected by steel tie wires, welding, or exothermic welding to create at least a twenty-foot (6.0 meter) length. Or, as an alternate method, a bare copper conductor not smaller than No. 4 AWG (21.15 mm$^2$), and at least 20 feet (6.0 meters) long. The rebar or copper wire must be encased in at least 2 inches (50mm) of concrete and extended horizontally near the bottom of the footing or foundation that is in direct contact with the earth.

The emphasis here is that the concrete must be in direct contact with the earth. And free from any plastic film, vapor barriers, or insulation.

And don't forget our discussion from Chapter 3 regarding a connection from this grounding electrode to a supplemental grounding electrode in order to protect the concrete from possible destruction due to ground-seeking currents (lightning).

With this additional protection, the concrete-encased electrode is an excellent system, and certainly inexpensive, as compared to a ground ring. This is due to the vast amount of reinforcing rod surrounded by a minimum of two inches (50 mm) of concrete, which is in direct contact with the earth. The resistance-to-ground of this system is very low, even in dry soil. In fact, it was first used in Tucson, Arizona. This area has a dry, sandy soil, where other types of grounding electrodes (ground rods) may not be acceptable.

As a brief history of the concrete-encased electrode, we go back to 1942, and to a munitions facility (bomb storage) at Davis-Monthan Air Force Base in Tucson, Arizona. The concern in this facility was the possibility of ignition caused by lightning and other sources of static electricity. The arid soil conditions of Tucson presented a problem for conventional grounding methods. Ground rods would not have been effective.

Mr. Herb Ufer was a consultant for the U.S. Army. And he decided that the reinforcing rods in the concrete footings could prove to be an effective grounding system. There were a total of 24 buildings, and Mr. Ufer performed ground resistance tests at each building from 1942 until 1960. During this period he determined, that at no time, did any resistance reading exceed five ohms. And the average resistance reading over this eighteen year period was 3.57 ohms. In 1960, he submitted these test results to IEEE, and this grounding electrode was initially referenced in the 1968 NEC. Since then, this grounding electrode has been known as a 'Ufer Ground'.

4. **Ground Ring**-This type of grounding electrode has gained favor in recent times, especially where sensitive electronic equipment may be affected by outside disturbances, including, but not limited to, lightning. And where there is a need to equalize the potential gradient around a building or structure.

5. **Rod and Pipe Electrodes**- This type of grounding electrode is the most common, and, unfortunately, the most widely abused. This is, very likely, caused by misconceptions of the proper use of this grounding electrode.

Let's examine the type of rods that are available, and their proper use.

The NEC identifies the conditions of use as follows:

Electrodes of pipe or conduit shall be at least 3/4 inch (metric designator 21), in diameter, and if made of steel, the outer surface must be galvanized, or otherwise metal-coated for corrosion protection.

Rod-type electrodes of stainless steel, copper, or zinc-coated steel must be 5/8 inch (15.87 mm) in diameter, unless listed (UL 467), whereby, they may be 1/2 inch (12.7 mm) in diameter.

The fact is that a galvanized rod has a significantly lower current carrying capacity than a copperbonded rod of the same diameter. For example, a 5/8 inch (15.87 mm) copperbonded rod is the approximate equivalent of a 3/0 AWG (85 mm$^2$) copper conductor, whereby, a 5/8 inch (15.87 mm$^2$) galvanized steel rod is the approximate equivalent of a No. 1 AWG (42.40 mm$^2$) copper conductor.

However, this is not the most important consideration. This is due to the fact that the resistance-to-ground of the single rod would typically be high enough to limit the current flow into the earth anyway.

The main issue is the service life of the rod. Depending on soil conditions (corrosivity), a galvanized only rod will have a service life of about ten years. Whereas, a copperbonded rod, which has an outer copper sheath, with a thickness of .25 mm, has an average service life of 30 years. Copperbonded rods are available with a copper sheath of .33 mm thick, and an average service life of 40 years.

A one-inch (25.4 mm) diameter copperbonded ground rod is the approximate equivalent of a 400 kcmil copper conductor (203 mm$^2$). And if the ground rod was one-inch galvanized steel, it would be the approximate equivalent of a 250 kcmil copper conductor (127 mm$^2$).

The equivalent continuous current ratings of these conductors, from NEC Table 310.15(B)(16) at 75°C., would be as follows:

5/8″ – (15.875 mm) – copperbonded rod 3/0 AWG – (85 mm$^2$) – 200 amperes

5/8″ – (15.875 mm) – galvanized steel rod – No. 1 AWG (42.40 mm$^2$) – 130 amperes

1″ – (25.40 mm) – copperbonded rod – 400 kcmil – (203 mm$^2$) – 335 amperes

1″ – (25.40 mm) – galvanized steel rod – 250 kcmil (127 mm$^2$) – 255 amperes

The first point is that we are not concerned about the continuous current ratings of these conductors, or the ground rod for that matter. These conductors and the ground rods may be carrying current at low levels, possibly, up to ten percent of the total current returning to the supply transformer, with the remaining 90 percent returning to the source through the supply grounded (neutral) conductor. Our concern is the short-time current rating of the conductors and associated ground rods. Lightning currents reach a peak in approximately two to ten microseconds. Now, let's examine the short-time ratings (fusing current) of these conductor sizes for 1/8 cycle (.002 seconds) as a comparison:

5/8″ – (15.875 mm) copperbonded rod – 3/0 (85 mm$^2$) – 1/8 cycle (.002 seconds) = 518,200 amperes (1083°C.)

5/8″ – (15.875 mm) galvanized steel rod – No.1 AWG (42.40 mm$^2$) – 1/8 cycle (.002 seconds) = 258,950 amperes (1083°C.)

1″ – (25.40 mm) copperbonded rod – 400 kcmil (203 mm$^2$) – 1/8 cycle (.002 seconds) = 1,235,350 amperes (1083°C.)

1″ – (25.40 mm) galvanized steel rod – 250 kcmil (127 mm$^2$) – 1/8 cycle (.002 seconds) = 772,100 amperes (1083°C.)

We can easily see that the short-time (1/8 cycle-.002 seconds) fusing current of these conductors is extremely high, as compared to their normal continuous current ratings from NEC Table 310.15 (B)(16), at 75°C. And lightning currents reach a peak much faster than 1/8 cycle (.002 seconds). So the fusing currents, or the amount of current that it would take to cause the conductor to fuse or melt in .002 seconds (1/8 cycle), is much lower than the fusing current for two to ten microseconds. Lightning currents, even those of many hundreds of thousands of amperes, would not even come close to damaging the ground rod, or the conductor attached to it.

It is easy to see that very large conductors attached to a ground rod would offer no additional protection.

To provide guidance in the understanding and application of conductor short-time current carrying capacity, based on insulation withstand ratings, fusing or melting currents for conductors, and termination integrity, where conductors terminate in switches, circuit breakers, controllers, etc., we have included Tables to identify these conditions of use. These Tables may be used in conjunction with the calculated fault currents to determine conductor insulation withstand ratings (150°C,) (NEC Section 110.10)), and, the amount of current and time to cause a conductor to fuse or melt (1083° C). as well as temperature limits at conductor terminations that may not affect the terminal integrity (250°C.).

In addition, chemically-charged ground rods are commonly used, and they may be quite effective, especially where soil conditions are unknown. The listing standard for this type of grounding electrode is UL 467J.

These electrodes have a hollow core, which is filled with a type of salt, possibly magnesium sulfate, which is designed to react with the normal moisture in the air, dissolve, and slowly migrate into the surrounding soil through weep holes in the wall of the rod. These rods are supplied with a low resistivity medium, which may be bentonite (natural clay), or another material of low resistivity which surrounds the outer surface of the rod. These electrodes have a service life of up to 50 years.

I have used these types of electrodes in several locations in Africa and three locations in North Korea. The latest installation was at a hospital in Ethiopia. A few days after the chem-rod was installed, I checked the resistance-to-ground and the measurement was 2 ohms. Of course, this reading will decrease over time as the salt from the rod slowly migrates into the soil.

These electrodes may be configured in straight, or 'L' shape designs. In Africa, because freezing conditions are not an issue, and the installation difficulty is greatly reduced, the resistance-to-ground results are similar to a straight rod.

6. **Plate Electrodes** - This electrode is not very commonly used, except where bedrock is a problem.

The NEC states that the plate electrode must expose at least two square feet (0.186 m²) of surface to exterior soil. Iron or steel plates must be at least ¼″ (6.4 mm) in thickness, and nonferrous plates must be at least .06″ (1.5 mm) in thickness.

The plate must be buried to a minimum depth of 30 inches (750 mm).

Generally, the plate is buried on edge, rather than laid flat

Where these buried plates are used as the grounding electrode for a lightning protection system the sides of the buried plate provide a means of capacitive coupling to the earth. The flat surfaces of the plate have a much lower impedance to the earth than a round ground rod when subjected to the relatively high frequency of lightning current. At higher frequencies, flat conductors have a lower impedance than round conductors, due to the 'skin effect' of alternating current.

For example, a 4/0 AWG copper conductor (107.21 mm$^2$) has an effective impedance of .074 ohms per 1000 feet (304.8 meters) at .85 power factor.

However, a 4/0 AWG copper conductor (107.21 mm$^2$) at 10 mHz, and only 10 feet long (3.048 meters), has an impedance of 232 ohms. Once again, this is due to the skin effect of alternating current at higher frequencies. At these higher frequencies, round conductors would not be effective as a bonding means. Also, it should be noted that above 100 Hz, inductive reactance becomes more important than resistance with respect to conductor impedance. So, due to the increased frequency of lightning, there is further evidence of the need to limit the length of grounding electrode conductors, as $XL = 2\pi FL$.

An interesting comparison can be made between the surface area of a 5/8 inch (15.875 mm) diameter by 10-foot (3 meters) long ground rod, and a one square foot (0.0929 square meter) steel or nonferrous plate. The surface contact area of this ground rod is 75 square inches (48,397 square mm, or 0.0484 square meters). And the surface contact area of the steel or nonferrous plate (2 sides) is 288 square inches (185,806 square mm, or 0.186 square meters). So, it is easy to see that the surface contact area of the plate is significantly greater than that of the rod. And, once again, because it is flat, as opposed to round, its impedance at higher frequencies (lightning) is much lower because of skin effect and the capacitive coupling effects to the earth.

But, the key element here is depth. The ground rod should reach a depth of, at least, 8 feet (2.44 meters), where possible, and the plate may be only 30 inches (750 mm) deep (250.53(H)). Of course, the plate may be placed deeper, where the soil conditions may be conducive to a lower resistance-to-ground electrode.

This completes our analysis of the various grounding electrodes that are referenced in Article 250 of the NEC. Further application information will be included in Chapter 8, as we address the grounding of systems and equipment.

# 5

# Bonding

A definition of this term 'Bond' or 'Bonding' appears in NEC Article 100, and it states 'connected to establish electrical continuity and conductivity'. A previous definition of this term stated, 'the permanent joining of metallic parts to form an electrically conductive path which will ensure electrical continuity, and the capacity to safely conduct any fault-current likely to be imposed'.

In either case, the concept is the same. Bonding may be achieved through the use of the various metallic raceways or cable assemblies that are recognized as equipment grounding conductors, as well as metallic cable tray as listed in NEC 250.118. And, this would include the appropriate fittings that are associated with these wiring methods.

Bonding accomplishes a means of equalizing voltage differences for electrical systems and equipment. And it is critical for the operation of normal functions, as well as those systems that are sensitive to even small external disturbances. Touch potential differences are certainly of paramount importance.

In this Chapter, we will examine the different methods and requirements of bonding, not only for 50/60-cycle systems, but also high frequency bonding techniques.

A definition appears in Section 250.2 which addresses a conductor that insures the bonding integrity of metallic enclosures on the supply side of a service, or for a separately-derived system. This conductor is identified as a 'Bonding Jumper, Supply-Side' and it derives its size from Section 250.102(C)(1) and (2).

Proper bonding at these locations, especially on the supply-side of the service equipment, is extremely important because the only protection ahead of this location is the overcurrent protection on the primary side of the utility transformer, which may be 3-6 or more times the full-load primary current rating. And, this condition is compounded further if the supply system is a Delta connected primary and a 3-Phase, 4-wire, Wye connected secondary, where a 30 degree phase-shift from primary to secondary causes the primary voltage to lead the secondary voltage by 30 degrees, and that may leave the secondary unprotected as well.

47

The general requirements for bonding are referenced in NEC 250.4(A)(3) and (4) for solidly grounded systems. Section (3) specifies the need for the connection of the noncurrent-carrying equipment which encloses electrical conductors or equipment in order to establish an effective ground-fault current path.

Section 250.4(A)(4) stresses the requirements for the proper bonding of the noncurrent-carrying electrically conductive materials that are likely to become energized, so that an effective ground-fault current path is established.

This effective ground-fault current path is referenced in 250.4(A)(5). This conducting path is required to carry the maximum ground-fault current, from the point where the ground-fault develops, to the source of the electrical supply (transformer, generator, etc.). The maximum ground-fault current must be capable of causing the prompt clearing of the appropriate overcurrent device on a solidly-grounded system.

If the system is high-impedance grounded, a ground detector must initiate the alarm or signaling system, which would identify a ground-fault (250.36 – 250.187).

Section 250.4(B)(2) applies to similar bonding requirements for an ungrounded system. For an ungrounded system, it is still important to limit the voltage-rise (touch potential) during a ground-fault.

In addition, the low-impedance grounding circuit would assure the prompt clearing of the overcurrent devices if a second ground-fault developed before the first ground-fault is cleared on an ungrounded system.

This applies to electrically conductive materials that are likely to become energized (metal raceways metallic enclosures, etc.) (250.4)(B)(3)).

And, as for solidly-grounded systems, a low-impedance grounding path must be established from the point of the ground-fault to the location of the grounding electrode conductor connection, and to the point of connection at the supply transformer, generator, or other supply source.

Whether the system is solidly-grounded, impedance-grounded, or ungrounded, continuity of the bonding or grounding path must be assured. And nonconductive coatings must not cause the interruption of the effective bonding or grounding path (250.12),(300.6(A)).

The next bonding connection is the Main Bonding Jumper, as defined in Article 100, (250.28). This is the connection between the grounded conductor at the service and the equipment grounding system. Of course, the vast majority of the ground-fault current must flow through this bonding connection.

With this in mind, we recognize that the Main Bonding Jumper is not required to be as large as the service supply conductors because it is designed to carry ground-fault current for only the very brief period until the overcurrent device of the faulted circuit, or possibly the service overcurrent device, promptly clears. So its physical size is typically smaller than the size of the ungrounded service conductors (Table 250.102(C)(1)). Once again, we are not concerned with its continuous current-rating, but its short-time current-carrying capacity.

I would like to recognize the System Bonding Jumper, which is also defined in Article 100. This conductor performs the same function as the Main Bonding Jumper. It is installed at the source of a separately-derived system (transformer, generator, etc.). Or, at the first disconnect or overcurrent device that is supplied by the separately-derived system (250.30(A)(1)). Its size, whether it is of the wire type or a busbar type, is based on Table 250.102(C)(1).

I would also like to mention that Section 250.30(A)(1), Exception No. 2 permits a System Bonding Jumper at both the source of a separately-derived system and the first disconnecting means. This exception applies where an outdoor transformer or generator supplies the building or structure. However, if this exception is used, there cannot be a conducting parallel path for the grounded (neutral) conductor. This conducting path could be a metallic raceway, or an equipment grounding conductor. So, to use this exception, the supply conductors for the building or structure would be in a nonmetallic raceway(s), or a nonmetallic cable(s), with no equipment grounding conductor run with the supply conductors.

Just as in the case of the Main Bonding Jumper, the System Bonding Jumper is designed to carry ground-fault current only for the time that it takes for the faulted circuit overcurrent device to clear. So, its continuous current rating is not the issue. The short-time current carrying capacity of this conducting path is relative to its physical size.

Finally, where the size of the service conductors, or the feeder conductors of a separately-derived system are larger than 1100 kcmil (558 mm$^2$) copper or 1750 kcmil (887 mm$^2$) aluminum, the Main Bonding Jumper, the System Bonding Jumper, and the Supply-Side Bonding Jumper, derive their size on a factor of 12½% of the area of the largest ungrounded supply conductor, or equivalent for those conductors installed on parallel. (250.102(C)(1), Note 1).

## Load-Side Bonding Jumpers

NEC 250.102(D) specifies that equipment bonding jumpers on the load side of overcurrent devices are to be sized in accordance with the minimum equipment grounding conductor sizes listed in 250.122.

Section 250.102 (E)(1) states that where this bonding conductor is within a raceway, it may be bare, or, where insulated, it is to be green in color, or green with one, or more, yellow stripes (250.119). And, the continuity of the bonding path must be assured, and not affected by the removal of a device or a luminaire (250.148 (B)).

Section 250.102(E)(2) permits the equipment bonding jumper to be on the outside of a raceway or enclosure, if the length of the external bonding jumper is limited to 6 feet (1.8 meters), and the bonding jumper is routed with the raceway or enclosure.

It should be noted that, according to listing instructions, flexible metal conduct and liquidtight flexible metal conduct do not provide an effective grounding path in lengths exceeding 6 feet (250.118(5),(6)),(ANSI/UL1).

It is important to note that when grounding conductors or bonding conductors are installed on the outside of metal raceways or enclosures, their impedance is increased. In fact, IEEE Paper No.54 states that when this is done, the impedance of the grounding or bonding conductor is twice what it would be if it were inside the raceway or enclosure. This is the reasoning behind Section 300.3(B), which states that all conductors of the same circuit (AC), including equipment grounding and bonding conductors, shall be contained within the same raceway, auxiliary gutter, cable tray, cablebus assembly, trench, cable, or cord. A similar rule for conductors installed underground is listed in Section 300.5(I), where all of the conductors of the same circuit (AC) are required to be in the same raceway or cable, or installed in close proximity in the same trench.

Once again, these requirements are associated with the important capacitive coupling effects of the conductors, as well as reducing inductive heating and limiting circuit impedance.

In effect, magnetic-flux density decreases in accordance with the square of the distance between conductors of the same circuit.

Installing bonding jumpers on the outside of raceways for a limited length of 6 feet (1.8 meters), as permitted in 250.102(E)(2), is acceptable because the increase in circuit impedance for this short length is negligible.

Of course, the potential for physical damage to the external bonding jumper must be a consideration. However, in a Hazardous (Classified) Location, Sections 501.30(A),(B), 502.30(A),(B), and 503.30(A),(B) address the grounding and bonding requirements. Where flexible metal conduit and liquidtight flexible metal conduit is used an external bonding jumper, not exceeding 6 feet (1.8 meters) in length, may be used in accordance with Section 250.102(E). While external bonding jumpers are always subject to physical damage, and visual inspection will identify broken or damaged connections, this is most important in Hazardous (Classified) Locations, because bonding integrity must be assured in order to eliminate the potential of arcing across discontinuous sections of the wiring system which may lead to ignition of the atmosphere.

It should be noted that Section 250.102(E)(2), Exception recognizes an external bonding jumper, or supply-side bonding jumper longer than 6 feet (1.8 meters). This may be used for outside pole locations. And this would provide a means of bonding or grounding isolated sections of metal raceways or elbows in exposed risers, and for bonding grounding electrodes.

Another important bonding function is from Sections 250.50 and 250.58, which require a common grounding electrode or system. A building or structure may be supplied by separate services (230.2), or more than one feeder or branch circuit (225.30). And, where these supply systems are required to be connected to a grounding electrode, the same grounding electrode must be used. Two or more grounding electrodes that are <u>bonded together</u> are considered as a single grounding electrode system in this sense.

The bonding of the grounding electrodes of different systems serves the purpose of limiting the potential (voltage) differences between these systems, and any wiring system that is a part of these systems. This includes the bonding requirement for the grounding electrode of a lightning protection system (250.60),(250.106).

In addition, the grounding electrode(s) installed for Communications Systems (800.100(D), Television and Radio Equipment (810.21(J), CATV Systems (820.100(D)), and Network-Powered Broadband Communications Systems (830.100(A)), are required to be bonded to all grounding electrodes of other systems.

However, there is an exception to this bonding concept, where the normal bonding of grounding electrodes is prohibited.

This involves portable or mobile equipment operating at over 1000 volts. Portable is identified as equipment that may be easily carried from place to place. And mobile, which is defined as the ability to be moved freely or easily, possibly on wheels, tracks, or by similar means.

The supply system for this type of equipment will have a neutral conductor that is grounded through an impedance device (resistor). Where a Delta system is used to supply the portable or mobile equipment, a system neutral point and neutral conductor must be established and installed, such as through the use of a grounding transformer (zigzag autotransformer).

The grounding electrode for this portable or mobile equipment supply system must be isolated and separated in the ground from other grounding electrodes of the other systems by at least 20 feet (6 meters). There can be no metallic connection between this isolated grounding electrode, including any underground metallic piping system (250.188(E)).

The isolation is meant to prevent a voltage-rise on the frames of this equipment, which may occur if this portable or mobile equipment had grounding systems bonded to other electrodes.

The general rules of Sections 250.50, 250.58, and 250.60, 250.106 for lightning protection systems, require that the grounding electrodes of different systems be bonded together to form the grounding electrode system. This bonding requirement serves two purposes. The first is to reduce the effects of voltage differences between separate systems, For example, the grounding electrode (system) for an alternating current system (250.24(A)) bonded to the grounding electrode (system) for a communications system (800.100(D)).

The second purpose, and certainly an added benefit, is the reduction of the resistance-to-ground of the grounding electrode system.

In this example, Section 800.100(D) requires that the bonding conductor between the two grounding electrodes must be at least No. 6 AWG copper (13.3 mm$^2$), or equivalent. Of course, both systems may be connected to the same grounding electrode.

However, sometimes auxiliary grounding electrodes (250.54) may be used to supplement the equipment grounding system with no physical bonding conductor connecting the auxiliary electrode to any other grounding electrode. If used as a supplement for the equipment grounding conductor, this conductor provides the bonding means. For example, a branch circuit supplies a motor, and this circuit includes the ungrounded conductors and the equipment grounding conductor. Due to an excessive circuit length, a decision is made to connect the motor frame to a ground rod as a supplement to the equipment grounding conductor. This may be done, even though the equipment grounding conductor may have been increased in size to satisfy the provisions of Section 250.122(B), in order to compensate for voltage-drop.

As a modification to the general rules of bonding and grounding from Article 250, when dealing with Information Technology Equipment (Chapter 6 - Article 645), Sections 645.14 and 645.15 apply to System Grounding, Equipment Grounding, and Bonding. Remember, Section 90.3, the Code Arrangement, states that Chapters 1 through 4 apply generally. And Chapters 5, 6, and 7 apply to special equipment, occupancies, or other special conditions. Article 645 (Chapter 6), and specifically Section 645.14, covers System Grounding for Information Technology Equipment.

This section states that power supplies that are derived within listed Information Technology Equipment that supply information technology systems through receptacles or cable assemblies supplied as part of this equipment, shall not be considered separately-derived for the purpose of applying 250.30. This means that the grounding provisions of separately-derived systems, as specified in 250.30, may not apply.

For example, an isolation transformer that is a part of listed Information Technology Equipment may have a secondary connection to an auxiliary grounding electrode (250.54), and the normal required bonding of grounding electrodes (250.50 and 250.58) would not apply. This does not mean that this is an unsafe condition, because the auxiliary grounding electrode is connected to the equipment grounding system of the power supply (isolation transformer). And the equipment grounding conductor of the supply circuit is connected at the source (or further upstream) to the grounding electrode for the separately-derived system, or service-supply system, where the circuit originates.

So, in effect, this is the same as the previous example of the motor circuit.

The connection of the listed Information Technology Equipment to an auxiliary grounding electrode (250.54) may serve as an aid to hold this equipment at earth potential, but the use of this ground connection as the sole means of system and equipment grounding is strictly prohibited.

The instructions provided by the manufacturer of the listed Information Technology Equipment, as well as product standards for this equipment, must be closely followed to assure that the grounding and bonding provisions of Article 250 have not been compromised.

NEC Section 250.92 identifies the bonding requirements for Services. In this case, the bonding of Service Equipment and associated wiring systems is critical because there is, virtually, no overcurrent protection on the supply side of the service disconnect, except for the overcurrent protection on the primary side of the utility transformer.

This is where the use of bonding type locknuts and/or grounding type bushings are required to be used to assure the continuity of metallic-type wiring methods, especially where there are concentric or eccentric knockouts, or where reducing washers are used. However, even on clean knockouts, this type of bonding is required.

Of course, it is within the service equipment enclosure, and where the Main Bonding Jumper is installed, that the grounding electrode conductor extends to the grounding electrode. Although the grounding electrode conductor may extend to the grounding electrode ahead of the service equipment.

Section 250.94(A),(B) requires that a means be provided at the service equipment or metering equipment for intersystem bonding purposes. This external bonding system provides a set of terminals to accommodate the bonding of other systems. This includes Conductive Optical Fiber Cables (770.100(B)(2)), Communications Systems (800.100(B)(2)), TV and Radio

Equipment, (810.21(F)(2)), CATV Systems (820.100(B)(2)), and Network-Powered Broadband Communications Systems (830.100(B)(2)).

The fact is that by providing a means of intersystem bonding, external to the service equipment, such as a listed intersystem bonding terminal device, will serve to limit the voltage differences between these systems. This is a critical part of any electrical installation, due to the limited withstand rating of the equipment. Proper bonding methods, coupled with a grounding electrode (system) that has a low resistance-to-ground, as well as the use of surge-protective devices, will provide the best means of protection for all of these systems.

On the load side of the service disconnect, Section 250.96 requires the bonding of the metallic wiring methods that make up the equipment grounding system. This provision applies, whether or not, these metallic wiring systems are supplemented by equipment grounding conductors.

It should be noted that there is a connection between 250.96(A) and 300.6(A), which applies to protection against corrosion and deterioration. Certainly, it is critical to consider the continuity of threaded raceways, especially where these metallic raceways are threaded in the field and are to be installed in a corrosive environment. An approved <u>electrically conductive compound</u> must be used to coat the threads to protect against corrosion and to assure the continuity of the metallic raceway system. Once again, this protection requirement applies, whether or not, the metal raceway is supplemented with an internal equipment grounding conductor.

For circuit operating at over 250 volts-to-ground, the bonding integrity of metal raceways, and metallic cable assemblies for other than service conductors, must be assured by compliance with Section 250.92(B) as follows:

   a. Threaded couplings or threaded hubs on enclosures that are made wrenchtight.
   b. Threadless couplings and connectors made tight for raceways and metallic cables.
   c. Bonding-type locknuts, bushings, or bonding-type bushings.

Also, the exception of 250.97 is commonly applied, which includes double locknuts on clean knockouts. If the enclosure, with concentric or eccentric knockouts, is listed for use without additional bonding requirements, then bonding jumpers are not necessary.

NEC Section 250.96(B) applies to the use of Isolated Grounding Circuits, where an insulated equipment grounding conductor is run with the circuit conductors and is used for the purpose of grounding electronic equipment. The circuit would typically originate at the building service equipment and terminate at the electronic equipment. A metal raceway which contains this circuit would be isolated from the equipment enclosure through the use of listed nonmetallic fittings. This isolated metal raceway must be grounded, even though it is not used as the equipment grounding conductor.

This method of isolation is used for equipment that is hard-wired.

For cord-and-plug connected equipment, Section 250.146(D) applies to Isolated Ground Receptacles. These devices have been available for many years. Once again, an insulated equipment grounding conductor would be run with the circuit conductors, and it would terminate on the receptacle grounding terminal, which is isolated from the receptacle mounting strap or yoke. A second equipment grounding conductor is provided to ground the receptacle mounting strap and the metal box for the receptacle.

This second equipment grounding conductor may be a metal raceway or a listed metallic cable assembly (250.118).

I would like to add that I have been aware of these methods of isolation for many years. I have always questioned the validity of this form of isolation because 1 know of no research that has proven how effective this may be. It may prove to be hazardous if the equipment grounding conductor were extended to a dedicated grounding electrode, so that this grounding system has its own ground reference point, independent of the grounding electrode (system) for the building. This would be a direct violation of 250.50, and 250.58, and a potentially dangerous safety hazard.

These concepts of isolating the grounding conductor to reduce, or possibly eliminate, electromagnetic interference on the grounding system have been acceptable for many years. The mere fact that the equipment grounding conductor is insulated, and it may pass through upstream subpanels, so as to terminate at the equipment ground terminal bus within the service equipment, may provide the desired reduction or elimination of electromagnetic interference.

It should be noted that 'electrical noise' is generally caused by the operation of equipment on the premises. This includes loads such as, arc welders, electric arc furnaces, motors, motor control relays, and other control relays that produce a rapid succession of transient voltages on the normal voltage waveform, resulting in harmonic currents of relatively high frequencies. Sensitive circuits, including information technology systems, communications circuits, data processing systems, and instrumentation circuits, should be physically isolated from these 'noise' producing systems.

In addition, a common source of electromagnetic interference is related to loose or improperly tightened terminations. This problem tends to worsen over time as the contact resistance increases, resulting in higher temperatures.

Proper torquing of the terminals according to the manufacturer's instructions is imperative and a requirement of 110.3(B). It is a good idea, where possible, to mark across the terminal with a marker which will indicate a loosening of the connection through a visual inspection. If retightening is necessary, the terminal should be torqued to 90% of the original listed specification.

As a further explanation, where conductors of AC systems are installed within the same raceway or cable assembly (300.3(B), there will be voltage differences between the equipment grounding system and the grounded (neutral) conductor. And this condition is unavoidable. These voltage differences are the result of normal capacitive coupling between the ungrounded, grounded, and equipment grounding conductors. The capacitive-charging current in the equipment grounding conductor will be relatively low. And the resulting voltage difference between this equipment grounding conductor and the grounded (neutral) conductor may be considered objectionable for electronic equipment. There may be data errors or equipment damage. This problem has caused many designers and installers to attempt to isolate the electronic equipment from the normal grounding system. One method of isolation has been through the use of a separate and dedicated grounding electrode (system). Or, by the removal of the bonding connection between the grounding system for the electronic equipment and the rest of the electrical system. This is _never_ an acceptable remedy. In fact, these alterations may produce even greater disturbances. And, more importantly, personal safety will be compromised. Even with a dedicated grounding electrode (system), there will be voltage differences between this isolated grounding system and the other grounding electrodes for the electrical service, separately-derived system(s), and lightning protection systems, which have been effectively bonded to satisfy Sections 250.50, 250.58, 250.60, and 250.106. And, in addition, the required bonding of the grounding electrodes for Communications Systems (800.100(D), Radio and TV equipment (810.21J), CATV Systems (820.100(D), and Network-Powered Broadband Communications Systems (830.100)) may be at significant voltage differences from the isolated grounding system, especially during lightning activity.

And, even if the electronic-equipment is connected to a separate grounding electrode, this electrode is not totally isolated from the others because the earth, between the electrodes, is a conductor.

Finally, I am aware of recent studies that serve a purpose in the potential reduction of electromagnetic interference and the effects of harmonic currents. The use of steel raceways, rigid metal conduit, intermediate metal conduit, and electrical metallic tubing, may reduce the effects of EMI by as much as 95 percent, thereby negating the need to isolate the metal raceway from the equipment frame by the use of the listed nonmetallic fittings. This may be an effective solution, and a means of reducing or eliminating electromagnetic interference.

## Bonding Metal Water Pipes and Exposed Structural Steel

Section 250.104(A) requires the bonding of metal water piping systems installed in or attached to a building or structure to the service equipment enclosure, the grounded conductor at the service, the grounding electrode conductor, or to one or more grounding electrodes used for the service.

Once again, this bonding connection is designed to keep the water piping system at the same potential as the service equipment, with the exception of the voltage-drop in the bonding conductor.

In multiple occupancy buildings, it is possible to have metal water piping in each occupancy isolated from other occupancies through the use of nonmetallic water pipe. In this case, a bonding jumper may be connected to the metal water pipe and connected to the equipment grounding terminal of the switchboard or panelboard that serves that occupancy. Its size is determined by the rating of the main overcurrent device supplying the occupancy, and a reference to 250.102(D) and 250.122. This requirement is addressed in 250.104(A)(2).

Where exposed structural metal is connected to form a metal building frame, and it has not been intentionally grounded, and the metal frame is likely to become energized, a bonding connection must be extended to the service equipment enclosure, the grounded conductor at the service, or to the disconnect for the building or structure which is supplied by a feeder or branch circuit, or to the grounding electrode conductor, or to one or more of the grounding electrodes used for the building. Its size is determined by Table 250.102(C)(1), based on the size of the ungrounded conductors supplying the building (250.104(C)).

Metal building frames can become energized by lightning, or where electrical equipment is mounted to the metal frame of the building for support.

Section 250.104(B) stresses the need of bonding other metallic piping systems, including gas piping that may become energized, to the service equipment enclosure, the grounded conductor at the service, the grounding electrode conductor, or to one or more of the grounding electrodes in use, or the equipment grounding conductor of the circuit that may energize the piping system.

For example, a gas furnace may have its metal frame energized by the failure of electrical insulation of a furnace component. Or, possibly a solenoid valve may fail due to a faulted electrical supply circuit and energize the metal pipe in the process, or possibly insulation failure inside a gas range. Generally, this equipment is grounded through the equipment grounding conductor of the supply circuit. And, this would provide a means of grounding the metal gas pipe. And, this is not a violation of 250.52(B),(1), which prohibits the use of a metal underground gas pipe as a grounding electrode. Typically, the serving gas supplier installs an insulating coupling to isolate the interior gas pipe from the underground system.

In more recent Code cycles, it has become a requirement to bond metal water piping systems and exposed structural metal that is formed into a building frame to the grounded conductor of <u>each</u> separately-derived system. The connection to the metal water pipe is made to the nearest available point of the metal water piping system that is served by each separately-derived system (250.104(D)).

The size of the bonding conductor for the metal water pipe and the exposed structural metal frame is determined by Table 250.102(C)(1), and based on the largest ungrounded conductor of the separately-derived system.

These bonding provisions do not apply if the metal water piping system or structural metal is used as the grounding electrode for the separately-derived system (250.104(D)(2), Exceptions No.1 and No.2).

The metal water piping system may by bonded to the structural metal frame, and a single grounding electrode conductor would be extended to the grounded conductor of the separately-derived system from the metal water piping system, or from the structural metal frame .

Section 250.194 is a Section of the NEC that mirrors the provisions of IEEE 80-2000-Guide for Safety in AC Substation Grounding. This Section covers the Grounding and Bonding of Fences and Other Metal Structures.

In this type of electrical environment, it is critical to consider the effects of step, touch, and transferred potential. A metal fence surrounding a substation or switchyard is similar to the core of a transformer. And the collapsing magnetic fields from the conductors and equipment within the substation or switchyard will induce a circulating current in the fence fabric and anything connected to the fence (fence posts, barbed wire, etc.).

The substation will have a buried ground mat, which will typically consist of an arrangement of copper conductors, usually 4/0 AWG (107 $mm^2$) copper, or larger, on 10-20 foot centers (30.48-60.96 meters) and buried 18 to 24 inches deep (45.72-60.96 centimeters).

A layer of large stones, with sharp projections, are arranged on the ground, and above the buried ground mat. These large stones provide an additional layer of resistance. And this serves to reduce the level of step and touch potential, in that the sharp projections of the large stones limit the surface contact area on the bottom of the feet of an individual walking or working in this area, as compared to walking or standing directly on the soil.

Equipment and structural steel within the substation are bonded to the buried ground mat. Large equipment, such as transformers, will typically have two connections to the ground mat, arranged on opposite corners of the equipment (250.194(B)).

In addition, prefabricated wire mesh is available. It is designed to be placed below the large stone layer and above the buried ground mat. This is especially important in those areas where people may be operating equipment within the substation. The wire mesh would be bonded to the ground mat and to the equipment frame or steel support to limit the effects of touch and step-potential.

Metal fences that are within 16 feet (5 meters) of exposed conductors or equipment are bonded to the ground mat (250.194(A)). This bonding is done at the fence corners, and at intervals of 160 feet (50 meters) along the fence. Where overhead lines pass over the fence, a bonding connection is required on each side of the crossing. A bonding connection is made at gate support posts, and the gates are to be bonded to their support posts. A buried bonding jumper is to be installed at any fence opening, such as at the gates, to assure electrical continuity. This buried jumper is connected to the fence gate posts.

Barbed wire installed above the fence fabric is bonded to the fence, and this may be accomplished through the use of listed split-bolt connectors.

The ground mat must be extended beyond the swing of the gates. This is meant to protect a person approaching from the outside from the effects of touch and step-potential.

Whether to extend the ground mat beyond the periphery of the fence, or even a buried conductor beyond the fence line to provide protection for persons approaching from the outside is a decision of the designer. This may not be possible due to property constraints. However, if this is done, the mat or buried conductor is usually extended a distance of 3 feet (1 meter) outside of the fence. If the buried conductor is used and not the ground mat, it should be bonded to the fence at the corners, and, at least as often as the fence fabric is bonded to the ground mat. The burial depth of this outside conductor is 18 inches (0.5 meter). It should be noted that very high voltage may be present on the metal fence due to the collapsing magnetic field that is associated with the medium and high voltage electrical systems within the substation or switch area.

If this cable cannot be buried three feet (1 meter) beyond the fence line, the next best method would be to bury it beneath the fence line. While not offering the same level of protection from step and touch-potential as extending the cable three feet (1 meter) beyond the fence, some protection is provided by this method.

The physical size of this conductor is the same size as that used for the ground mat.

Another important concern is the possibility of transferring potential through the fence to a remote area. If this is of concern, especially where the fence may extend to a common area, and accessible to the general public, the continuity of the metal fence must be interrupted by a 10 foot (30.48 meters) section of wood fence, or other effective means.

## Bonding in Hazardous (Classified) Locations

Certainly, the bonding integrity between metal raceways, metallic cable assemblies, and metal enclosures must be assured to prevent arcing at joints and terminations. Even relatively low energy arcs or sparks may be a source of ignition. Indeed, even static electricity is of serious concern.

Fuel/air mixtures that are heavier than normal air, and that are within their ignitable range, may be ignited by thermite reactions, such as a ferrous tool striking a concrete floor.

Chapter 5 of the NEC has many sections that relate to proper bonding. Specifically, Sections 501.30(A), 502.30(A), 503.30(A), 505.25(A), and 506.25(A) deal with bonding requirements in Class I, Class II, Class III, Zone 0, 1, 2, and Zone 20, 21, 22.

Generally, the normal double-locknut and locknut-bushing type of connections are not acceptable for bonding purposes. Properly sized bonding jumpers are required. This bonding provision is required to the point on the supply system where the Main Bonding Jumper is installed within the service equipment. Or, where the System Bonding Jumper is installed at the point of grounding at a separately-derived system.

We have mentioned that Section 240.102(E)(2) permits an equipment bonding jumper on the outside of a raceway or enclosure. The length of the bonding jumper is limited to 6 feet (1.8 meters), and it must be routed with the raceway or enclosure. This applies to Hazardous (Classified) Locations, as well (501.30(A)), (502.30(A)), (503.30(A)). In these special locations, it may be beneficial to use an external bonding jumper so that a visual inspection may detect a broken connection.

Flexible metal conduit and liquidtight flexible metal conduit do not provide an effective grounding path in lengths exceeding 6 feet (1.83 meters), in accordance with the listing of these flexible raceways (ANSI/UL1).

## Health Care Facilities

And now the bonding and grounding in Critical Care Areas of a health care facility, specifically, a patient-care vicinity, where a patient equipment grounding point is established. This equipment grounding point would consist of one or more listed grounding and bonding jacks. A No. 10 AWG (5.26 mm$^2$) conductor is used to connect the grounding terminal of all grounding-type receptacles to the patient equipment grounding point. This bonding conductor may be looped or in a central location.

The installation of the listed grounding and bonding jacks in the patient-care vicinity for the purpose of reducing touch-potential differences is optional, but certainly an important consideration (517.19(D))

In the Patient-Care Vicinity, the grounding and bonding integrity must be assured in accordance with 517.19(E). For the case where feeder conductors are installed in metal raceways, or Type MC or MI cable that have been identified as an equipment grounding conductor according to 250.118, the grounding of panelboards, switchboards, and switchgear must be assured in a similar manner as service equipment. Grounding bushings with properly sized bonding jumpers (250.122) are to be used with metal raceways. Or, threaded hubs may be used on enclosures for the metal raceways or connectors for the cables. Bonding-type locknuts are also acceptable to assure the grounding integrity, and this provision excludes the use of standard locknuts for bonding purposes. In addition, the bonding jumpers are sized in accordance with 250.122 (517.19(E)(1),(2)).

## Swimming Pools, Fountains, and Similar Installations

The final item on our list and summary of the concept of bonding, and the one that seems to tie everything together in proper order, is the equipotential bonding requirements of a swimming pool in Section 680.26. Voltage differences in the immediate vicinity of a pool are a primary concern. Proper bonding methods will reduce these voltage differences (gradients) to acceptable levels.

Section 680.26(B) identifies the parts of the pool that are required to be bonded through the use of a solid copper conductor, not smaller than No. 8 AWG (8.37 mm$^2$). This conductor may be insulated, covered, or bare.

The conductor is buried and in contact with concrete, so corrosion is an important consideration. Solid conductors resist corrosion far better than stranded wires because they expose less of their cross-sectional area than a stranded wire, and using an insulated conductor as a means of corrosion protection is certainly an important consideration.

Rigid metal conduit may be used where it is of brass construction or identified as corrosion-resistant metal. This rigid metal conduit is typically used to connect a pool deck box to a wet-niche luminaire.

Pressure connectors that are listed as suitable for grounding and bonding equipment (250.8(A)) must be used to make connections to bonded parts.

The conductive forming shell of the pool serves as the common bonding grid (680.26(B)(1)).

However, where the reinforcing steel is encapsulated for corrosion protection, a copper conductor grid is required.

This copper conductor grid consists of No. 8 AWG (8.37 mm$^2$), or larger, bare solid copper conductors, properly bonded at all cross junctions with approved pressure-type connectors (250.8(A)).

The grid must conform to the contour of the pool.

It must be arranged on a 12 inch by 12 inch (300 mm by 300 mm) network in a uniformly spaced grid pattern.

And it must be secured within or under the pool, no more than 6 inches (150 mm) from the outer contour of the pool shell (680.26(B)(1)).

All metallic parts of the pool structure, including the following must be bonded (680.26 (B)(3), (4), (5), (6), (7)).

a. Metal forming shells and mounting brackets of no-niche luminaires
b. Metal fittings within or attached to the pool structure
c. Metal parts of electrical equipment associated with the pool water circulating system
d. All fixed metal parts, including metal raceways, metal cable sheaths, other metal piping systems, and metal fences

It is interesting to note that in the rare case where none of the bonded parts (metal forming shells of underwater lighting, metal ladders, metal diving board stands, etc.) are in direct contact with the pool water, an approved corrosion-resistant conductive surface that exposes at least 9 square inches (5800 mm$^2$) of surface area to the pool water at all times is required (680.26(C)) for bonding purposes (680.26(B)).

Once again, the bonding system is for the purpose of reducing the voltage gradient in the pool area. Voltage-drop in the bonding system will effect these voltage differences. Just as voltage-drop in equipment grounding conductors produce voltage differences in the grounding system.

But, it is important to note that the purpose of bonding is different than the purpose of grounding. While it is true that equipment that is bonded is also grounded, with certain exceptions, such as equipment that may be isolated by elevation (250.110)(1), and that is more than 8 feet vertically (2.44m), and more than 5 feet horizontally (1.52m) from ground or grounded metal objects, and free from contact by people. The purpose of bonding equipment and systems, as well as the bonding of grounding electrodes (250.50 - 250.58), is to reduce voltage differences. While the purpose of grounding is to reference systems and equipment to the earth, or to a conducting body that serves in place of the earth (vehicle frame, frame of an airplane, frame of a portable or vehicle-mounted generator, etc.). This connection establishes a '0' volts reference point.

It is not the intention of Section 680.26 to connect the bonding conductor for the pool to any grounding electrode (as some local authorities have required). Or, to extend the bonding conductor to any subpanel or service panel for the purpose of grounding. The bonding conductor serves no useful purpose in the opening of any circuit overcurrent device during a ground-fault.

The bonding grid (reinforcing steel, copper bonding grid, etc.) may not be used as a grounding electrode or as a part of the grounding electrode system. A significant voltage-rise may be present during lightning activity or from a ground-fault, and this would be hazardous to anyone in the vicinity of the pool (250.52(B),(3)).

# 6

# Ground-Fault Protection of Equipment

It started in 1971, with the addition of NEC Section 230.95. Ground-Fault Protection of Equipment is required for three-phase, four-wire, solidly-grounded, Wye electrical systems, with a voltage of over 150 volts, phase-to-ground, but not exceeding 1000 volts, phase-to-phase, where the rating of the service disconnect is 1000 amperes or more. The typical voltage for this system is 480/277 V. This type of system is capable of sustaining an arcing ground-fault. And, due to the extreme heat energy of this fault, significant damage may be the result. This is to say nothing of the arc-flash hazards to people.

Since 1971, additional references have been added, and these include NEC Sections 215.10, 240.13, 517.17, 695.6(G), 700.27, 701.26, and 708.52.

The original Section 230.95 identified this requirement for Services. Section 215.10 expanded this requirement for Feeders.

Section 240.13 was added to clarify that Ground-Fault Protection applies to a building or structure disconnecting means on this same type of supply system, whether this disconnecting means was classified as a Service Disconnect or a Feeder Disconnect. This Section also waives this requirement for certain continuous industrial processes, where a nonorderly shutdown may increase hazards to people or equipment. Or, where ground-fault protection is provided by other requirements for services or feeders. This may involve an installation where a downstream feeder is afforded this protection through an upstream service disconnect.

Also, Section 695.6(G) prohibits ground-fault protection for fire pumps. Indeed, the overcurrent protection for fire pumps is required to be able to carry, indefinitely, the locked-rotor current of the largest motor and the pressure maintenance pump motor and any additional motor load associated with this function. This is so critical, that where this calculation results in a value that does not correspond to the standard overcurrent device ratings in Section 240.6(A), the next larger standard device rating must be used (Section 695.4(B),(C)).

65

Keep in mind that Section 695.6(B)(1) requires that conductors supplying a fire pump motor(s) must be sized at not less than 125 percent of the sum of the fire pump motor(s) and pressure maintenance motors, plus 100% of the associated fire pump accessory equipment. This may seem excessive, in that the overcurrent device may be rated significantly above the ampere rating of the motor(s) circuit conductors. But we know that this is typical for normal motor circuits. For example, Section 430.22, where motor circuit conductors have an ampacity of 125% of a motor full-load current rating.

And Table 430.52 permits the maximum rating or setting of the motor branch-circuit overcurrent device to be above the ampacity of the motor circuit conductors. Also, Section 430.52(C)(1) has exceptions which permit even higher ratings or settings of the overcurrent devices.

The required method of connection to a listed fire pump controller is a direct connection to the power source (695.4(A)).

Remember, in our discussion of conductor insulation short-circuit current ratings, where we analyzed the 5 second, 1 second, 1 cycle, 1/2 cycle, 1/4 cycle, and 1/8 cycle ampere ratings of typical 75° C. rated insulation. These conductors are capable of carrying significant levels of current, for short periods of time, with no insulation damage. This is the reasoning behind these conditions. As far as the fire pump system, the concern is not equipment or conductor damage, but to extinguish the fire.

Getting back to our analysis of Ground-Fault Protection, Section 517.17 requires this protection for Health Care Facilities. In this application, where Ground-Fault Protection is provided at the service disconnecting means, an additional level of ground-fault protection is <u>required</u> for the feeder disconnecting means downstream and toward the load (Section 517.17(B)).

It makes sense that selective coordination between the pick-up setting of the downstream GFP be arranged to operate before the pickup setting of the main GFP at the service disconnect (Section 517.17(C)). Performance testing is a requirement to assure proper operation when this system is first installed (Section 517.17(D)).

Sections 700.31 and 701.26 apply to Ground-Fault Protection for Emergency Systems and Legally-Required Standby Systems. However, these systems are not required to have an automatic-trip function for the main disconnecting means. But, instead, an audible and visual alarm system must be provided to indicate a ground-fault. The sensor for the signal device is to be located at, or ahead of, the main system disconnecting means for the Emergency or Legally-Required Standby System. The maximum setting of the signal device shall be for a ground-fault of 1200 amperes (Sections 700.6(D) and 701.6(D)).

In addition, 700.32 requires that emergency system overcurrent devices be selectively coordinated with all supply-side overcurrent protective devices.

And 701.27 requires the selective coordination of legally-required standby system overcurrent devices with all supply-side overcurrent devices.

Article 705 includes the requirements for Interconnected Electric Power Production Sources. This may include a typical utility supply system and an on-site generator, a solar photovoltaic system, or a wind-powered system.

Ground-Fault Protection of Equipment may not be necessary for an interactive system. But, if used, the output of this system must be connected to the supply side of the GFP (Section (705.32).

Article 708 addresses Critical Operations Power Systems (COPS). These systems are identified by municipal, state, federal, or other standards such as NFPA 1600-2013. Supply and distribution systems include fire alarm, security systems, communications systems, power systems, HVAC systems, and signaling systems for critical operations areas.

Due to the critical nature of these systems, and their direct relationship to life safety, as well as continuity of business functions, and national security, selective coordination of critical operations power systems overcurrent devices must be assured with all supply-side overcurrent protective devices (708.54). In addition, Ground-Fault Protection of Equipment is required for these critical systems (Section 708.52). This requirement is the same as that for Health Care Facilities. Where the GFP is provided for the service or feeder disconnecting means in accordance with Sections 230.95 or 215.10, an additional level of GFP is required at the next level of feeder disconnecting means downstream toward the load. Selectivity is required to assure that the downstream protection operates without affecting the main GFP (Section 708.52(D)). And, the same requirement for testing the GFP functions that applies to Health Care Facilities, also applies to Critical Operations Power Systems. When the GFP is first installed, each level of ground-fault protection must be tested to be sure that it is operational.

Now, let's examine the need for this protection. In an earlier section of this book, we stated that the majority of faults occurring in electrical distribution systems are ground-faults. In a utility study of their distribution systems, it was determined that 90% of faults are ground-faults. And, 90% of ground-faults are arcing-type faults. These arcing ground-faults may produce currents that are in the range of 43% to 70% of bolted short-circuits. Where these arcing ground-faults continue, and the upstream overcurrent protective devices do not promptly clear (if they do, at all), the arcing fault causes considerable damage. This may, and typically does, result in a bolted short-circuit. This is the equivalent of two ungrounded conductors that are bolted together. The bolted short-circuit should cause the overcurrent protective device to promptly clear. But, by this time, the damage has been done. Another consideration in determining the impedance of the ground-fault current path (the equipment grounding system) is that the arcing ground-fault will have an associated voltage-drop, and this compounds the problem, as the level of ground-fault current is reduced. Unfortunately this may, and probably will, have an effect on the operation of the overcurrent protective device.

In consideration of this reduced ground-fault current flow, the equipment grounding system is a first line of defense in protecting systems and equipment from the potential devastating effects of arcing ground-faults. This impedance is of paramount importance in returning the ground-fault current to the system source in such magnitude that the appropriate overcurrent device will promptly clear.

This is easier said than done. The equipment grounding system may consist of any of the conducting paths that are referenced in Section 250.118. Or, combinations of these items, all the way back to the service equipment, or to the source of a separately-derived system. Or, to the first system disconnect or overcurrent device that is supplied by the separately-derived system (Section 250.30(A)(1). Either the Main Bonding Jumper 250.28(A),(B),(C),(D), in the service equipment, or the System Bonding Jumper for the separately-derived system, will permit the ground-fault current to return to the source. And when this circuit is complete, the ground-fault must be promptly eliminated.

Let's analyze a 400 ampere circuit on a 208/120 volt, 3-phase, 4-wire, Wye connected system, protected by a 400 ampere, molded-case circuit breaker. The conductors are installed in a 3 inch rigid metal conduit with threaded connections (wrench tight). We are looking for a ground-fault equivalent to 5 times the rating of the 400 ampere circuit breaker, or 2000 amperes. At this level of ground-fault current, in this example, it has been determined that this circuit breaker will clear in one second. We factored a 50 volt-drop into our equation as a consideration. In this case, the voltage becomes 70 volts, and at a current of 2000 amperes, the circuit impedance is .035 ohms.

$$\frac{70\,volts}{2000\,amps} = .035\,ohms$$

The impedance of a 3 inch rigid metal conduit at 2000 amperes for a length of 1000 feet is .099 ohms in our example. If we reduce this to a 100 foot length, it is .0099 ohms. These values are for a rigid metal conduit (3″) without couplings. Even if the conduit connections were made wrenchtight, there will be a significant increase in the conduit impedance. So, we increase the conduit impedance by 50 percent.

$$\frac{.0099\,ohms \quad (100\,Feet, 3"\,RMC, without\,couplings\,at\,2000\,amperes)}{\times 1.5}$$

$$.01485\,ohms \quad (100\,Feet, 3"\,RMC, with\,couplings\,at\,2000\,amperes)$$

We can now determine the length of this conduit by dividing the circuit impedance (.035 ohms), by the impedance of the 3 inch rigid metal conduit with threaded (wrenchtight) coupling connections.

$$\frac{.035\ ohms}{.01485\,ohms} = 2.35, \text{ or 235 feet, or 71.628 meters}$$

To summarize this example, we have deduced that, in this circuit, a 400 ampere circuit breaker will clear this arcing ground-fault in one second, at a current of 2000 amperes. We factored an arcing ground-fault with a voltage-drop of 50 volts. The system voltage is 208/120 (3-phase, 4-wire, Wye connected system). The circuit conductors are installed in a 3 inch rigid metal conduit with wrenchtight threaded connections. At a ground-fault current of 2000 amperes, and a conduit impedance of .01485 ohms, as well as a length of 235 feet (71.628 meters), this 400 ampere circuit breaker cleared the arcing ground-fault in one second.

If it is necessary to increase the length of this system beyond 235 feet (71.628 meters), the conduit may be supplemented by an internal equipment grounding conductor (NEC Section 300.3(B)). Based on the rating of the 400 ampere circuit breaker, and a reference to NEC Table 250.122, the <u>minimum</u> size of copper equipment grounding conductor is No. 3 AWG (26.67 mm$^2$). If we use the conductor impedance values from NEC-Chapter 9, (Effective Impedance at 0.85 Power factor for uncoated copper wire per 1000 feet (304.8 meters), the impedance is 0.24 ohms. Reducing the impedance to 100 feet (30.48 meters) equals .024 ohms. Combining this ohmic value with the impedance of the rigid metal conduit (with wrenchtight couplings) for 100 feet, and factoring the result into our equation, we can determine the increase in length.

$$\frac{1}{\frac{1}{.01485} + \frac{1}{.024}} = 0.00917 \text{ ohms (conduit/conductor impedance)}$$

$$\frac{.035\,ohm\ (circuit\,impedance)}{.00917\,ohms} = 3.817, or\,382\,feet\ (116.43\,meters)$$

We have increased the length of the conduit from 235 feet (71.628 meters) to 382 feet (116.43 meters) by the use of the No. 3 AWG copper equipment grounding conductor in parallel with the 3″ rigid metal conduit.

This is a good example for training purposes. And, based on the given parameters, we have determined that if a ground-fault occurred anywhere in this circuit, the 400 ampere circuit breaker would open or clear in one second.

I am not advocating that a one second clearing time is acceptable. Due to the heat energy of a 2000 ampere ground-fault for one second, serious damage may be the result. If a faster clearing time is required, that would have to be factored into the equation.

For example, if a one-cycle clearing time was necessary (.0167 seconds), the ground-fault current would have to be significantly increased, depending on the time-current curves of the circuit breaker (or fuse).

The rigid metal conduit impedance at the higher current level would be determined, and then the calculation would dictate the conduit length.

In examining some test data from a report for the same 3 inch rigid metal (steel) conduit, and the same circuit parameters (208/120 volt system, 400 ampere circuit, 50 volts-drop in the arcing ground-fault), the ground-fault current of Modeling and Testing of Steel, IMC, and Rigid (GRC) Conduit, Part I, May 1994, Copyright 1994, Georgia Tech Research Institute, comparisons may be made and data from studies such as this may be used to determine the validity of a metal conduit as an equipment grounding conductor. Aluminum conduits, with an oxide inhibitor at the conduit joints, can be extended for greater distances, due to their nonmagnetic properties.

And now, let's complete our summary of Ground-Fault Protection of Equipment.

We identified the NEC requirements for this type of protection. In effect, solidly-grounded, 3-Phase, 4 Wire, Wye electrical systems of more than 150 volts-to-ground, and not exceeding 1000 volts, phase to phase, where the rating of the Service or Feeder disconnect is 1000 amperes, or more.

This device is equipped with relays that are set to recognize the presence of a ground-fault. And if the fault-current reaches the set point of the relay, a switch or circuit breaker will open and clear the ground-fault.

The ground-fault protection relay does not, in itself, provide the ground-fault protection. And the NEC permits a ground-fault current to reach as high as 1200 amperes before the GFPR initiates the opening of the disconnect switch or circuit breaker. A time-delay of one second is permitted for ground-faults of 3000 amperes, or more.

The pickup setting of the GFPR should be in conjunction with the fuse(s) or circuit breaker(s) that actually provides the overcurrent protection. This means that criteria used in determining the setting of the ground-fault protection relay must be arranged to allow the appropriate fuses and circuit breakers to perform their intended function, before the GFPR signals the main disconnect to open.

In some cases, reports have been noted that a ground-fault on a 20 ampere branch-circuit has caused the GFPR to initiate an entire system black-out. The setting must allow the downstream overcurrent devices to operate.

Where selective coordination is required, this becomes extremely important. NEC Sections 517.17 and 708.52 require 100 percent selectivity. This means that the overcurrent devices for the ungrounded conductors must be able to recognize all levels of ground-fault current, regardless of how high or low this current may be. So, the protection afforded by the ground-fault protection is the last resort for these critical systems.

As I write this summary and watch CNN, I am stunned by a news report from Atlanta, Georgia about a worldwide power outage of a major airline, whereby their entire system was shutdown. Evidently, this outage occurred at 2:30 AM (EDT) on 08/08/16. Some service was resumed by 9:00 AM (EDT). I could not help but wonder whether this system outage was caused by the operation of a ground-fault relay, or, the opening of a fuse or circuit breaker due to the lack of selective coordination of the overcurrent devices.

In some cases, there may be an arcing-fault burndown, where conductors and equipment actually melt. Such a condition would take a significant length of time to repair, on even a temporary basis.

One level, or step of ground-fault protection is usually sufficient. However, in special situations, two levels or steps are required.

As we have already stated, for health-care facilities, NEC Section 517.17(B) identifies where ground-fault protection must be provided on all of the downstream feeders toward the load. This downstream protection does not apply to the 'Essential Electrical System'.

NEC Section 517.17(C) requires that this protection be fully selective, so that the feeder device opens a ground-fault on the load side of the device without affecting the GFP at the service. No separation of the time-current characteristics of these devices is specified (for example, a minimum of six cycles). But only that manufacturers recommendations be followed to achieve 100 percent selectivity.

NEC Section 708.52(B), (C), and (D) requires this two-level, or two-step protection for Critical Operations Power Systems.

Once again, 100 percent selectivity is required between the ground-fault protection at the service disconnect and the downstream ground-fault protection at the next level of all feeder disconnects toward the load.

Certainly, ground-fault protection has been an NEC requirement since 1971. On 480/277 volt systems that are solidly-grounded, and that have service disconnects of 1000 amperes and larger, there is enough arcing-fault energy to sustain an arcing ground-fault for longer periods. We know that arcing faults may produce temperatures of 35,000 degrees F., or 19,427 degrees C. This heat energy will vaporize conductors and equipment. To say nothing of the damage to people at these extremely high temperatures. This heat energy is about four times the surface temperature of the sun.

Of course, there have been instances where the pick-up setting of the ground-fault relay has been set too low. And the time-delay setting may also be set for an instantaneous trip, and this may result in an unnecessary outage.

Unfortunately, this has caused some people to disconnect the ground-fault protection.

Due to the need of selective coordination, whether this is required, or not, the benefit of having ground-fault protection, in conjunction with properly applied overcurrent protection cannot be overemphasized.

Possibly, the use of inverse-time ground-fault relays, which afford a practical means of adjusting the time-delay feature to a value that eliminates an unnecessary trip function, as well as setting the ampere set-point as high as may be necessary, may allow the system overcurrent devices to perform their intended purpose, and eliminate unnecessary power failures.

Maybe, it would be beneficial to design supply systems that would eliminate the need for ground-fault protection.

An example would be to supply large systems through the use of multiple service disconnects with smaller overcurrent devices. For many years, installations have been supplied by several service disconnects, as opposed to one. For example, using 6 - 800 ampere disconnects (NEC Section 230.71), grouped at the same location (NEC Section 230.72), instead of a single service disconnect.

Possibly, the choice of a high-resistance grounded 480 Volt, 3-Phase, Wye connected system, in lieu of a solidly-grounded system would eliminate the need for a ground-fault protection system (NEC Section 250.36). Neutral connected loads may be derived from downstream transformers at 480/277 volts, and less than 1000 amperes. Or, another alternative is to use 208/120 volt systems instead of 480/277 volt systems. On another note, as a means to distribute the load into smaller systems, I have seen a suggestion by some people in the industry that makes sense when alternate power sources are used. Especially for health care facilities, where the normal and emergency supply systems may be quite large. It is common to see the use of one large transfer switch, where a more practical approach may be to use smaller transfer switches, strategically located on the premises wiring system, so as to be closer to the connected load. This may afford a greater degree of reliability for the connected load.

For example, in a health care facility in Ethiopia where I worked a few years ago, a 400 ampere transfer switch failed after only 3 years of use. Now, I must admit that the conditions there are extreme as compared to installations in the United States. The utility provides very unstable power, and, I have witnessed outages that number as many as 10-15 in one day. The transfer switch, which was of high quality, was opening and closing under significant load, and the contacts of one-phase (Phase C) failed on the utility side of the switch. The load on this system was not balanced, and Phase C was the most heavily loaded. So, it made sense that this was the phase that failed.

It should be noted that the normal and emergency systems were not only supplying the critical or essential hospital load, but the entire load. This involved the administration building and the housing load, which was significant.

This is a prime example where the installation of smaller transfer switches, near the connected load that they serve, would have enhanced the reliability of this system, as well as lengthening the service life of these transfer switches.

Another hazard that would have been eliminated is that when the 400 ampere transfer switch failed on the utility side, the back-up generator was forced to run 24 hours a day for a period of 10 days. If the generator failed, and all power was lost, this would have proved to be a serious disaster. This hospital is a six-hour drive from the capital city of Addis Ababa, where the generator dealer is located. A replacement switch had to be delivered from Europe (London). Hence, the 10-day period to replace the transfer switch.

However, the recommendations to use smaller and strategically located transfer switches is still a very sound consideration.

Now, this may be an extreme case in point. Certainly, facilities in the U.S. are not subject to the power problems of Ethiopia. So transfer switches are, for the most part, only operated once a week, as the generator is exercised to meet legal requirements. Or, during brief outage periods.

# 7

# Overcurrent Protection-Current Limitation Selective-Coordination

In this chapter we will address the requirements of these three topics.

To start, we will examine the definitions of the terms that apply to conditions produced by overcurrents.

**Current-Limiting Overcurrent Protective Device** - This device is designed to significantly reduce the current-flow in a faulted circuit to a level that is substantially less than the fault-current that would flow in the same circuit if this device were replaced with a solid conductor having comparable impedance. These devices, whether in the form of a fuse or circuit breaker, are designed to open a short-circuit in 1/2 cycle, or less (.008 seconds).

It makes a big difference as to the type and characteristics of the overcurrent device that is used for a particular application.

The overcurrent devices in use today are truly remarkable devices.

The use of current-limiting overcurrent devices affect the size of circuit components, including equipment grounding conductors. However, the use of these devices has to be carefully considered, because, where they may be used to reduce the available short-circuit current on their load side, selective coordination must also be considered.

These devices are designed to open when the current reaches, or exceeds, a specific level to reduce the effects of thermal stress in conductors and equipment. These thermal effects may be caused by overloads, short-circuits, or ground-faults.

An **Overload** is defined as the operation of equipment in excess of its full-load current rating, or a conductor which is operated above its rated ampacity, which over time, produces higher than normal temperatures in the equipment, or the conductor.

A **Short-Circuit** occurs due to a conducting connection between two ungrounded conductors, or, between an ungrounded conductor and the grounded conductor of a system.

A **Ground-Fault** occurs due to a conducting connection between an ungrounded conductor and the equipment grounding system.

75

**Overcurrent** - A current in excess of normal conditions, or above the ampacity of a conductor. The cause of the overcurrent may be the result of an overload, short-circuit, or ground-fault.

**Overcurrent Protective Device, Branch-Circuit** - This device is intended to provide overcurrent protection for services, feeders, and branch-circuits, as well as for equipment. This protection may range from its ampere rating to its interrupting rating.

**Overcurrent Protective Device, Supplementary** - These devices provide limited protection for utilization equipment, such as lighting units or appliances. They are not considered a substitute for the branch-circuit overcurrent device (240.10). And, they do not have a short-circuit interrupting-rating (UL1077).

**Circuit Breaker** - This device is meant to open and close by nonautomatic means. And to automatically open an overcurrent without damage.

**Fuse-** An overcurrent device with an element that is designed to open or part when subjected to a current flow above its rating.

**Short-Circuit Current Rating** - The symmetrical fault current that conductors and equipment may safely withstand without damage (110.10).

**Interrupting Rating** - The highest current at rated voltage that a device is intended to interrupt under standard test conditions. The highest current that must be interrupted is not always associated with short-circuit or ground-fault conditions. The interrupting rating, or horsepower rating of motor-circuit switches and motor controllers are examples of equipment that is designed to open under locked-rotor current conditions developed by motors without damage. Molded-case circuit breakers and molded-case switches are tested under overload conditions at six times their rating to assure their ability to interrupt locked-rotor currents (430.83(A)(1), 430.109(A)(1),(2),(3)).

In order to understand the concepts of overcurrent protection, including the application of all of these definitions, an actual analysis of a supply system, including a service supply from a utility transformer, a feeder, and a branch-circuit is necessary. This will also include the supply from a back-up generator. This installation is a healthcare facility in Addis Ababa, Ethiopia but, this installation can be anywhere, U.S.A., for that matter.

When we consider the use of Current-Limiting Overcurrent Protective Devices, we must emphasize the importance of the use of these devices. The physical size and overall cost of an electrical installation may be significantly reduced.

But, one word of caution that is an important consideration, is that these devices should be, and in some cases, must be, selectively coordinated with overcurrent devices that are on their load side.

For example, an 800 ampere current-limiting fuse is installed ahead of a 200 ampere noncurrent-limiting fuse. This noncurrent-limiting fuse has a one-cycle (.0167 seconds) clearing time, in its interrupting range. If a short-circuit fault occurs downstream of the 200 ampere fuse, and the short-circuit current is within the current-limiting range of the 800 ampere fuse, it may, and very likely will clear faster than the 200 ampere fuse. This affects the entire system. In addition, after the fault has been cleared, the effects on the integrity of the 200 ampere fuse may cause a failure at a later time.

Selective coordination should not be compromised, even if it is not required.

**Selective Coordination** is defined as the localization of an overcurrent condition to restrict outages to the circuit or equipment affected, accomplished by the selection and installation of overcurrent protective devices, and their ratings or settings for the full-range of available overcurrent conditions, that may range from an overload to the maximum available fault-current, and the full-range of overcurrent protective device opening times associated with these overcurrents.

In effect, we are trying to localize a fault so that the entire system is not disconnected. Most people, if not all people, have witnessed the effects of a complete system 'blackout'.

I distinctly remember the catastrophic power outage that affected parts of Ontario in Canada, as well as Connecticut, Massachusetts, New Hampshire, New Jersey, New York, Rhode Island, Vermont, and Pennsylvania on November 9, 1965. More than 30 million people were left without power, covering an area of 80,000 square miles (207,000 km$^2$) for up to 13 hours.

The cause of this power failure was traced to a <u>protective relay</u> on a transmission line which was improperly set. This protective relay, which was to trip if the current exceeded the capacity of the transmission line, was set too low. And the cascading effect of this failure caused major improvements to be designed for the entire grid.

Some alert load dispatchers at utilities in Connecticut, Massachusetts, New York, and New Jersey, disconnected their power plants from the grid and were able to supply power.

This is a perfect example of what we have been covering in our discussion of ground-fault protection and selective coordination.

There have been major power outages since then (New York, 1977). And due to the complexity of the power grid, these events will not be uncommon.

I think that it would be safe to say that many existing power distribution systems that are a part of premises wiring systems, do not have proper selective coordination in their design.

For many years, a fuse manufacturer has published selectively ratios based on fuse ratings (1.5-1, 2-1, 3-1, 6-1, etc), which are based on the line-side vs. load-side ratings of the fuses.

Circuit breaker manufacturers provide coordination tables that may be used in this process.

Various computer programs are available that allow a designer to select time-current curves from manufacturers and plot these curves on one graph. However, selective coordination is only achieved by a careful analysis of the time-current curves and their relationship to the available fault-currents in the system.

In any event, this type of protection is sometimes a recommendation. NEC Section 240.12 stresses that where an orderly shutdown is required to minimize hazards to people and equipment, a system of coordination based on the following two conditions shall be permitted:

1. Coordinated short-circuit protection
2. Overload indication based on monitoring systems or devices. An Informational Note states that a monitoring system may initiate an alarm, as opposed to an outage, which would allow corrective action, or an orderly shutdown that would minimize the hazard to people and equipment

For feeders and branch circuits of over 1000 volts, NEC Section 240.100(C) requires that the available short-circuit current and the selected conductor shall be coordinated to prevent damage or dangerous overheating in conductors or conductor insulation.

NEC Section 645.27 requires the selective coordination of the overcurrent devices for critical operations data systems with all supply-side overcurrent devices. These are information technology equipment systems that require continuous operation because of public safety, national security, emergency management, or business continuity. Selective coordination of overcurrent devices is extremely critical in these special systems.

NEC Section 700.32 requires overcurrent devices on emergency systems to be selectively coordinated with all supply-side overcurrent devices. In this case, the selective coordination must be documented and made available to those authorized to design, install, inspect, maintain, or operate the system.

NEC Section 701.27 requires the selective coordination of the overcurrent devices of Legally-Required Standby Systems with all supply-side overcurrent devices.

Legally-Required Standby Systems provide electric power for operation of fire-fighting, rescue operations, and the control of health hazards. This is in difference to the life safety functions that are addressed for Emergency Systems in NEC Article 700. Another difference from Emergency Systems is that when power is lost, it may be restored within 60 seconds, as opposed to within 10 seconds for Emergency Systems (701.12),(700.12).

NEC Section 708.54 requires selective coordination of the overcurrent device for critical operations power systems with all supply-side overcurrent protective devices.

The last concept of selective coordination involves the Ground-Fault Protection of Equipment requirements of NEC Sections 215.10 (Feeders), 230.95 (Service), 517.17 (Health Care Facilities), and 708.52 (Critical Operations Power Systems). The pickup-setting of the ground-fault protection should be, and in some cases, must be, selectively coordinated with downstream overcurrent devices.

My point is simply that ground-fault current must be determined by calculation and coordinated with the operating characteristics of the circuit overcurrent device and the impedance of the effective ground–fault current path. Which, according to NEC Section 250.4(A)(5), is to be of sufficiently low-impedance to facilitate the operation of the appropriate overcurrent device. It must be capable of safely carrying the maximum ground-fault current that may be imposed on it at any point where a ground-fault may occur back to the electrical supply source. If the supply system is not solidly grounded, the fault-current would initiate the operation of the ground detector for a high-impedance grounded system or an ungrounded system, and an alarm would identify the presence of a ground-fault (250.36(2)), (250.21(B)).

## Electrical Safety in the Workplace NFPA 70E

The last part of analysis regarding selective coordination and ground-fault protection involves the issue of electrical safety in the workplace.

We can readily understand that reducing an arc-flash hazard through the use of current-limiting overcurrent protective devices will enhance worker safety. And this is a hallmark of the purpose of NFPA 70E.

Let's go back to the beginning. On March 15, 1972, OSHA requirements became effective. There was no electrical consensus standard available at that time. The safety standard that had been in use since March 15, 1897 was, what we know today, as the National Electrical Code. The NEC, although recognized as a means of safeguarding persons and property from the hazards arising from the use of electricity (NEC Section 90.1(A)), it does not address the issues associated with workplace safety. At first, OSHA adopted certain NEC sections that would improve worker safety. These references appear in

OSHA 29 CFR, Part 1910, Subpart S, Part I. And, 1910.399 has definitions of electrical terms that were taken from NEC Article 100.

In 1976, an electrical standards development committee was formed. And their duty was to prepare a consensus standard that we know today as NFPA 70E. The first edition was published in 1979. A main concern has been the effects of arc-flash hazards. An arc-flash is defined in Article 100 of NFPA 70E as a dangerous condition associated with the possible release of energy caused by an electric arc.

It should be noted that an arcing-fault hazard, which may be line-to-line, line-to-neutral, or line-to-ground, produces both toxic and conducting gases. These gases may lead to an even greater amount of destructive energy to which a worker may be exposed.

Arc-flash hazards exist whenever conductors or circuit parts are exposed, and not in an electrically safe work condition. This means that they have not been deenergized.

In most cases, systems can be deenergized before a person works on, or even near, energized parts. And this is an OSHA requirement in 1910.333(a)(1).

However, it is recognized that sometimes it may be hazardous to deenergize certain systems. Examples would be where deenergizing introduces increased hazards, such as disconnecting a ventilation system(s) for a Hazardous (Classified) Location. Or, the need to test voltages for diagnostic purposes.

NFPA 70E requires that a 'Shock Hazard Analysis' be done (Section 130.4(A)) to determine the level of voltage that a person may be exposed to, as well as the dimensions of the arc-flash boundaries, and the Personal Protection Equipment (PPE) that is required to reduce the effects of electric shock.

The conditions of maintenance and supervision play an important role in reducing arcing-fault hazards. This would include the maintenance of overcurrent protective devices. Infrared thermography is commonly used to identify problems of overheating at terminations, possibly caused by loose connections, improperly sized conductors, or internal damage to overcurrent devices.

I must emphasize that there is no substitute for visual inspections.

A comprehensive maintenance document that I have used extensively since 1970 is NFPA 70B - Recommended Practice for Electrical Equipment Maintenance. This book will serve as an invaluable aid in complying with the maintenance requirements of NFPA 70E.

And finally, NEC Section 110.16 requires a label to be affixed to switchboards, switchgear, panelboards, industrial control panels, meter socket enclosures, and motor control centers in other dwelling occupancies, which would identify that an arc-flash hazard may exist where this equipment is likely to require examination, adjustment, servicing, or maintenance while energized. This label may be factory or field-marked on the equipment.

For field-applied hazard markings, an Informational Note of Section 110.21(B) references ANSI Z535.4 - 2011 Product Safety Signs and Labels for further information.

NFPA 70E, Section 130.5(C), provides additional requirements relating to the identification of arc-flash hazards which are more specific than the general rule of NEC Section 110.16.

In my opinion, the most effective method of reducing the clearing time of a fuse or circuit breaker for the purpose of arc energy reduction to comply with 240.87(B), is Provision 4 of 240.67 and 240.87(B).This is an energy-reducing active arc-flash mitigation system.

This type of protection permits a technician to set an energy-reducing maintenance switch in order to reduce the clearing time of the circuit breaker while performing maintenance with the person working within the arc-flash boundary of NFPA 70E. A power system analysis must be performed to determine the fault current that will flow through the circuit breaker associated with the arc-flash reduction maintenance system unit .Transient load current must be determined including motor inrush current and transformer inrush current.

The fault current will normally be arcing and not bolted fault current. A pickup setting for the arc flash reduction maintenance system will be below 75% of the calculated arcing current and above the total transient load current. A Table in IEEE STD 1584 TM-2002 and formulas are used to calculate arcing fault current.

Effective on January 1, 2020, where fuses are rated 1200 amperes or higher, the clearing time must not exceed 0.07 seconds at the available arcing current, or one of the protection techniques referenced in 240.67(B) shall be provided. This includes an energy reducing active arc flash mitigation system

## NFPA 70 E Article 100 Definitions

**Arc Flash Hazard** - a source of possible injury or damage to health associated with the release of energy caused by an electric arc.

This thermal hazard is associated with an energy level of 1.2 calories or more per square centimeter. This energy level will typically be sufficient to cause a second-degree burn. The consensus is that there is a probability that the employee will not suffer a permanent physical injury.

**Arc Flash Suit** - complete arc-rated clothing and equipment that covers the entire body, except for the hands and feet. An arc flash suit with a rating of 40 calories per square centimeter (40 cal/cm$^2$) is common. Protection must also be provided for the hands and feet.

**Arc Rating** - The value attributed to materials to describe their performance when exposed to an electric arc discharge. The arc rating is expressed in cal/cm$^2$ and it is derived from the determined value of the 'arc thermal performance value' (ATPV) or 'energy of breakpoint threshold' ($E_{BT}$), whichever has the lower value. The 'energy of breakpoint threshold' is related to the formation of one or more holes in the innermost layer of arc-rated material that would allow flame to pass through the material.

ATPV and $E_{BT}$ are defined in ASTM F1959/F1959 M.

Arc rated clothing (AR) is also flame resistant (FR). But flame resistant clothing may not provide adequate protection from an arc flash. So, only arc rated clothing is acceptable to provide protection where PPE is required.

**Boundary, Arc Flash** - This describes an approach limit from an arc source at which incident energy equals 1.2 cal/cm$^2$ for one second, which is sufficient to cause a second-degree burn on unprotected skin.

**Boundary, Limited Approach** - an approach limit at a distance from an exposed energized electrical conductor or circuit part within which a shock hazard exists.

The limited approach boundary is the approach limit for unqualified persons.

**Boundary, Restricted Approach** - an approach limit at a distance from an exposed energized electrical conductor or circuit part within which there is an increased likelihood of electric shock, due to electric arc-over combined with inadvertent movement. The arc flash boundary may be greater, less than, or equal to the restricted approach boundary. This boundary is the approach limit for qualified persons. When crossing this boundary, the qualified person must be protected with the proper PPE.

**ANSI/IEEE 1584** - is the 'Guide for Performing Arc-Flash Hazard Calculations' and this standard may be used to determine the arc flash hazard distance and the incident energy to which employees could be exposed.

**Qualified Person** - one who has demonstrated skills and knowledge related to the construction and operation of electrical equipment and installations and has received safety training to identify the hazards and reduce the associated risk.

This defined term was introduced in Article 100 of the National Electrical Code in 1971, and it simply stated: 'one familiar with the construction and operation of equipment and the hazards involved'. This definition in the present NEC is virtually the same as in Article 100 of NFPA 70E.

**Electrically Safe Work Condition** - a state in which an electrical conductor or circuit part has been disconnected from energized parts and locked /tagged in accordance with established standards and tested to verify the absence of voltage, and, if necessary (especially for medium and high voltage systems), temporarily grounded for personnel protection. Checking for the absence of voltage is an activity, that, in itself, is hazardous, and for this reason, suitable personnel protection equipment (PPE) is required.

**Risk assessment** – a process that identifies hazards and estimates the likelihood of occurrence of injury or damage to health, and determines if protective measures are required

The risk assessment procedure must identify the process to be used by the employee before the work is started and it must include:

1. Identify hazards
2. Assess risk
3. Implement risk control according to the hierarchy of risk control methods.

When properly enforced and followed by the employees, these guidelines will provide the means to assure safe work practices.

**Unqualified Person -** A person who is not a qualified person.

**Arc Flash Boundary -** The distance at which a person is likely to receive a second degree burn ($1.2$ cal/cm$^2$).

# 8

# Review of NEC

## Article 250 Grounding and Bonding Review

In Section 250.4(A)(5), we have the term Effective Ground-Fault Current Path, and this is the path for ground-fault current to flow back to the electrical supply source in order to promptly clear the overcurrent device protecting a circuit. This would be for a solidly-grounded system. In summary, this effective grounding path has three provisions.

__First__, it is permanent and continuous, so any discontinuity is not acceptable. While this seems obvious and not worthy of consideration, I have seen installations where rigid metal conduit was used as the sole grounding path and the threaded joints were coated with various materials that may have interrupted the continuity of the grounding path. Section 300.6(A) indicates that where the conduit is threaded in the field, the threads shall be coated with an <u>approved electrically conductive, corrosion-resistant compound</u>. We can see the need for this Section so that we may have a permanent and continuous grounding path.

__The Second Element__ of the effective grounding path is that it has sufficient capacity to safely carry any ground-fault current that may be imposed on it. Therefore, the ground-fault current must be known, as well as the operating characteristics of the overcurrent device protecting the circuit, and the short-time current carrying capacity of the conductor. If the equipment grounding conductor is in the form of an insulated copper conductor, then calculate the short-time current rating of the insulation. I would also use this same calculation if the conductor is not insulated.

__The Third Element__ of the effective grounding path is that it has a sufficiently low impedance to limit the voltage-to-ground during ground-faults and facilitate the operation of the overcurrent device in a solidly-grounded system. The length of the grounding conductor must be considered, due to the fact that excessive length may produce significant voltage-drop in the grounding conductor, and that would have an effect on the voltage-rise, above earth potential, on the equipment grounding system during a ground-fault. Also, please refer to 250.122 to identify the minimum size of the equipment grounding conductor where it is in the form of a copper conductor, an aluminum conductor, or a copper-clad aluminum conductor. If there are problems associated with objectionable currents over the grounding conductors or grounding paths, caused by multiple connections between grounded conductors and equipment grounding conductors, the solution would be to remove all of these except the Main Bonding Jumper (250.28) or System Bonding Jumper (250.30) at the service equipment or separately-derived system. Where isolation of objectionable DC ground currents caused by cathodic protection systems is required, a listed AC coupling/DC isolating device is permitted in the equipment grounding path to provide an effective return path for AC ground-fault current while blocking DC current (250.6(E)).

Grounding conductors and bonding jumpers may be connected by exothermic welding (which is a listed process), listed pressure connectors, listed clamps, or other listed means. Sheet metal screws are not to be used to connect equipment grounding conductors to enclosures (250.8(A)).

## Circuit and System Grounding - Section 250.20

AC Systems are required to be grounded in accordance with Sections 250.20(A) and 250.20(B). These sections would apply to systems of up to 1000 volts.

For AC circuits of less than 50 volts, the grounding requirements are as follows:

1. Where the low-voltage circuit is supplied through a transformer supplied at over 150 volts-to-ground.
2. Where supplied by transformers, if the transformer supply circuit is ungrounded.
3. Where installed as overhead conductors outside of buildings where the low-voltage circuit may be affected by lightning induced influences, or where the low-voltage circuit may make accidental contact with higher voltage circuits.

For AC circuits of 50 volts to 1000 volts supplying premise wiring systems, the grounding requirements are as follows:

1. Where the system is grounded so that the maximum voltage–to-ground on the ungrounded conductors does not exceed 150 volts. (Such as a 3-phase, 4-wire, Wye-connected system operating at 208/120 volts).
2. Where the system is 3-phase, 4-wire, Delta connected, in which the midpoint of one phase winding is used as a circuit conductor.

3. Where the system is 3-phase, 4-wire, Wye connected, in which the neutral is used as a circuit conductor.
4. High-Impedance grounded neutral systems

High-Impedance grounded neutral systems are referenced in 250.36 and 250.187. If we have a 3-phase, 3-wire, 480 volt, Wye-connected system (no neutral loads), and instead of making a solid connection to ground, the connection to ground is made through a high-resistance element. Such a system would function the same as an ungrounded system, in that a single ground-fault will not cause overcurrent devices to clear. This type of system would provide better protection from transient voltages (lightning, etc.) than an ungrounded system. For example, if the maximum ground-fault current produced in this system is 5 amperes, the resistance element in series with the grounding electrode conductor and the neutral conductor would be rated at 55.4 ohms.

$$\frac{277\,volts}{5\,amperes} = 55.4ohms$$

The neutral conductor from the neutral point of the Wye connection of the transformer to the resistor is fully insulated (600 volts), and is at least 8 AWG copper (8.37 mm$^2$) or 6 AWG aluminum or copper-clad aluminum (13.30 mm$^2$). There would be a voltage-rise above earth potential on this neutral conductor, and thus the reason for the conductor insulation.

For separately-derived systems as covered in 250.20(A) and (B), the grounding requirements are covered in 250.30.

Section 250.21 addresses AC systems of 50-1000 volts that are not required to be grounded.

1. The electrical systems used only to supply industrial electric furnaces are not required to be grounded, even though they operate in this voltage range
2. Separately-derived systems used only for rectifiers that supply only adjustable-speed industrial drives
3. Separately-derived systems that are supplied by transformers that have a primary voltage rating less than 1000 volts, if all of the following conditions are met:
    1. The system is used only for control circuits
    2. Qualified persons service the installation
    3. Continuity of control power is required
    4. Ground-detectors are installed on the control system

This provision would recognize that control circuits operating in this voltage range (50 v. – 1000 v.) may be ungrounded so that a ground-fault in the control circuit does not disconnect critical equipment. Also, a reference can be made to 430.44, which is associated with critical motor loads.

Section 250.24 has numerous subsections that apply to Grounding Service Supplied AC Systems.

In this instance, a connection to a grounding electrode is to be made from the grounded conductor. This connection may be made at any accessible point from the load end of the service- drop or service-lateral to and including the terminal or bus in the service equipment (250.24(A)(1)).

Where the transformer supplying the service is located outside of the building (typical), an additional grounding connection must be made from the grounded service conductor to a grounding electrode, either at the transformer, or elsewhere outside of the building. This connection to ground is mainly for lightning protection. This additional grounding connection does not apply to high-impedance grounded neutral systems (250.24(A)(2)).

If services are dual fed and employ a secondary tie, a single connection to ground may be made at the tie-point, instead of two grounding connections, one from each service (250.24(A)(3)).

If the Main Bonding Jumper (250.28) in the service equipment is in the form of a wire or busbar, connection to the grounding electrode may be made from either the grounded (neutral) conductor terminal bus or the equipment grounding conductor terminal bus (250.24(A)(4)).

Normally, when the grounded service conductor is connected to ground at the service equipment, an additional connection to ground from this grounded service conductor is not made beyond the service equipment. However, this does not apply to separately-derived systems (250.30), or for connections at separate buildings or structures (250.32).

The ground provisions for Separately-Derived Systems are referenced in 250.30. A System Bonding Jumper (250.30(A)(1)) shall be used to connect the grounded conductor to the equipment grounding conductor(s). This bonding jumper size is based on the cross-sectional area of the transformer secondary conductors, assuming the separately-derived system is a two-winding transformer supplying a premises wiring system. The bonding connection may be made at the transformer (source), or at the first disconnect or overcurrent device supplied by the transformer. If the grounding electrode to be used (250.30(A)(4)) is closer to the transformer (generator, solar photovoltaic system, or other supply source), than it is to the first disconnect or overcurrent device supplied from the source, then the connection to ground should be made at the source. If the grounding electrode is closer to the first disconnect or overcurrent device, the connection to ground should be made there. The reason for this provision is to limit the length of the grounding electrode conductor in order to hold the grounded conductor and the equipment grounding conductor(s) at, or near, earth potential ('0' volts). The greater the length of the grounding electrode conductor, the less effective it is in accomplishing this important condition. Also, due to the fact that one of the functions of the grounding electrode system is lightning protection, and because lightning has a relatively high frequency AC component, it would be important to install the grounding electrode conductor (250.66),(250.166) as straight as possible, avoiding unnecessary and especially sharp bends and limit the overall length, which would affect the impedance of the grounding electrode conductor (250.4(A)(1) Informational Note No.1). In addition, if physical protection is necessary for this conductor, use rigid nonmetallic conduit to eliminate any choking effects which would also increase the impedance of the grounding electrode conductor. If a steel enclosure (rigid metal conduit) is used for physical protection, make sure that the enclosure is bonded to the grounding electrode conductor at both ends in order to reduce the choking effects. If the physical protection for the grounding electrode conductor is in the form of a metal channel that does not completely encircle the grounding electrode conductor, there is no increase in impedance. This choking effect is due to induced current in the steel enclosure (250.64(B),(E)) (IEEE Paper 54). Nonferrous enclosures are not required to be bonded, due to their nonmagnetic properties (250.64(E)(1)).

The grounding electrode for the separately-derived system shall be the building or structure grounding electrode system. This will assure that the separately-derived system will have the same ground (earth) potential as other systems within the same building or structure.

If multiple separately-derived systems are installed, such as in a high-rise building, and these systems are connected to a common grounding electrode conductor, the common grounding electrode conductor is to be sized in accordance with 250.66, based on the total cross-sectional area of the largest derived phase-conductor from each separately-derived system. Taps from these multiple separately-derived systems to the common grounding electrode conductor shall be based on 250.30(A)(6), in accordance with the cross-sectional area of the ungrounded conductor(s) serving that separately-derived system. The connections to the common grounding electrode conductor are to be accessible and made by an irreversible listed compression connector or by the exothermic welding process (busbar connections are acceptable), and the common grounding electrode conductor is required to be not smaller than 3/0 copper (85 mm$^2$) or 250 kcmil aluminum (127 mm$^2$) (250.30(A)(6),(a)).

For Class 1, 2, or 3 circuits (Article 725), a grounding electrode conductor connected to a grounding electrode is not required where these circuits are supplied by a transformer rated not more than 1000 volt-amperes. The grounded conductor of this transformer would be connected to the grounded metal enclosure of the transformer (motor controller or disconnect), and the bonding jumper to make this connection is not to be smaller than No.14 AWG copper (2.08mm$^2$), or No.12 AWG aluminum or copper-clad aluminum (3.30mm$^2$). The enclosure for the motor controller or disconnect is to be grounded in accordance with 250.134. Of course, the enclosure frame is to be connected to the equipment grounding conductor (250.118) of the circuit supplying this equipment.

In addition, the exposed structural steel which forms the building frame, or the interior metal water piping, which is in the area served by the separately-derived system, is to be bonded to the grounding electrode conductor that is connected at the point where the system bonding jumper is connected, either at the source of the separately-derived system, or at the first disconnect or overcurrent device supplied by the separately-derived system (250.104(D)). Specifically, where the grounded conductor of each separately-derived system is bonded to the nearest available point of the interior metal water piping system in the area served by each separately-derived system 250.104(D)(1), the bonding jumper shall be sized in accordance with Table 250.102(C) (1), and the connection is to be made at the same point on the separately-derived system where the grounding electrode conductor is connected.

If exposed structural metal that is interconnected to form a building frame is in the area that is served by the separately-derived system, it must be bonded to the grounded conductor of each separately-derived system (250.104(D)(2)). Each bonding jumper is to be sized in accordance with Table 250.102(C)(1), based on the largest ungrounded conductor of the separately-derived system.

However, a separate bonding connection to the building structural metal is not necessary if the building steel is used as the grounding electrode for the separately-derived system (250.104(D) (2), Exception No.1).

And, if the metal water piping system is used as the grounding electrode for the separately-derived system, there is no need to provide a separate bonding connection to the water pipe (250.104(D)(1), Exception No. (1).

Finally, if the metal water pipe is used as the grounding electrode for the separately-derived system, the building structural metal may be bonded to the metal water pipe and a single grounding electrode conductor may be extended to the separately-derived system.

This bonding connection is to the made in the area that is served by the separately-derived system (250.104(D)(2), Exception No. 2).

## Two or More Buildings Supplied by A Feeder(s) or Branch Circuit(s)

Section-250.32 addresses a condition where two or more buildings or structures are supplied by a common service. This could involve dwelling occupancies, industrial, commercial, and institutional installations. There seems to be a lot of confusion regarding the grounding methods that are acceptable here. We have two choices to make in these cases, and it does not matter whether the installation is single-phase or three-phase. If one building or structure is supplied from another, the supply system may be made of an ungrounded conductor(s), a grounded conductor, and an equipment grounding conductor. In this case, the grounded conductor of the initial service is bonded to the equipment grounding conductor(s) at the initial Service Main Bonding Jumper (250.28),(250.102(C)(1)), and the supply system to the second building or structure consists of the ungrounded conductor(s), the grounded conductor, and the equipment grounding conductor.

The equipment grounding conductor is sized in accordance with 250.122.

Please do not forget that the equipment grounding conductor may be the metal raceway that surrounds and protects the supply conductors (250.118). Now, at the second building or structure, the grounded conductor is isolated from the equipment grounding system. If a connection is made between the grounded conductor and the equipment grounding system at the second building or structure, an objectionable current (neutral current) would flow through the equipment grounding conductor, eventually back to the initial service through the Main Bonding Jumper, and here this current will divide. Most of the current will return to the source through the grounded conductor of the service, but some of this current would flow through the grounding electrode conductor and into the earth, and then through the earth to the grounding electrode at the supply transformer, and then through the grounding conductor back to this transformer. I would estimate that 90-95 percent of this current would return to the supply transformer through the grounded service conductor, and 5-10 percent between the two grounding electrodes and the earth between them. This would be true for normal operating conditions and ground-fault conditions.

Therefore, in this case, there would not be a bonding connection between the grounded conductor and the equipment grounding conductor at the second building or structure (250.32(B)(1)). However, the equipment grounding system at the second building or structure will be connected to a grounding electrode(s) (250.50), (250.52), unless the provisions of 250.32(A) Exception apply. For example, this exception waives the requirement to connect the equipment grounding conductor to a grounding electrode at the second building or structure, where this building is supplied by a single branch-circuit. This Exception includes a multiwire branch-circuit.

In the <u>second case</u>, the supply system to the second building or structure has the ungrounded conductor(s), the grounded conductor, <u>but not the equipment grounding conductor</u>. Now, the grounded conductor is bonded to the building or structure disconnecting means and to the grounding electrode, and shall be used for grounding or bonding of equipment, structures, or frames required to grounded or bonded. The size of the grounded conductor is to be no smaller than the larger of:

1. That required by 220.61.
2. That required by 250.122.

The exception applies only to existing installations that are relevant to previous editions of this Code, in fact, prior to the 2008 edition of the NEC.

In using this second example, there can be no continuous metallic paths bonded to the grounding system in both buildings or structures, such as common water piping systems, (see the definition of the term Building in Article 100), and ground-fault protection of equipment (215.10, 230.95, 240.13, 517.17, 708.52) has not been installed on the common AC service where downstream grounding connections may have an effect on the operation of the ground-fault protection system.

This second example would not be acceptable in a location where neutral-to-earth voltages may be objectionable, such as in Agricultural Buildings (547.9(B)(3)), where even very small voltage differences are not acceptable.

If the system is ungrounded, the grounding electrode is connected to the equipment grounding system at the building or structure (250.32(C)).

Where one or more disconnecting means supply one or more additional buildings or structures that are under single management, and the disconnect(s) is remote from the building or structure in accordance with 225.32, Exception No.1, the connection of the grounded conductor to the raceway or metal-cable assembly (250.118) used for grounding any noncurrent carrying metal frames of equipment, interior metal piping systems, and building or structural metal frames, is not to be made. An equipment grounding conductor is to be run with the circuit conductors to the separate building or structure and bonded to existing grounding electrodes. Or, where there are no existing electrodes, the grounding electrode(s) required by Part III of Article 250 shall be installed, where a separate building or structure is supplied by more than one branch-circuit (250.32(A), Exception). The bonding of the equipment grounding conductor to the grounding electrode at the separate building or structure shall be made in a junction box, panelboard, or similar enclosure, which is located immediately inside or outside of the separate building or structure (250.32(D)).

The grounding electrode conductor shall be sized in accordance with 250.66, based on the largest ungrounded supply conductor (250.32(E)).

## Grounding Electrode System and Grounding Electrode Conductor

If available on the premises at each building or structure served, each of the grounding electrodes specified in 250.52(A)(1) through (A)(6) shall be bonded together to form the grounding electrode system (250.50, 250.58, 250.60, 250.106). It is very important to bond separate grounding electrodes in order to reduce the voltage differences which may appear from time to time. For example, if two different systems supply a building, such as the normal electric service and a communications (telephone) system, and each of these systems are connected to their own separate grounding electrode(s), a very serious problem could develop. Let's say that the grounding electrode for the electric service has a resistance-to-ground of 20 ohms, and there is an indirect lightning strike in the vicinity of the service transformer, and the lightning current reaches 15,000 amperes, there will be an instantaneous voltage-rise on the transformer secondary. Before this occurrence, each system was being held at earth potential ('0' volts). But now the lightning current being carried to ground through the service system is on the order of 15,000 amperes. With a 20 ohm resistance-to-ground, the instantaneous voltage-rise on the service system, above earth potential, will be 300,000 volts (20 ohms × 15,000 amperes). Meanwhile, the communications system is still at earth potential ('0' volts). The voltage difference between these two systems is 300,000 volts! By properly bonding these two grounding electrodes, the voltage difference may be several hundred or a few thousand volts, due to the inductive voltage-drop of the bonding conductor. There are many Sections in Chapter 2 and Chapter 8 that related to the bonding of separate grounding electrodes. For example, the bonding reference in 250.50. Also, 250.58 and 250.60, 250.106, as well as 800.100(D) for communications systems (also see Information Note No. 2 of this section), 810.21 (J) for

TV and Radio equipment, 820.100(D) for CATV systems, and 830.100(D) for Network-Powered Broadband Communications Systems.

Also, see 250.188(E) for the grounding of medium voltage systems supplying portable or mobile equipment (portable generators, cranes, shovels, dredges, etc). This is the only reference that requires the isolation of the grounding electrode for this equipment (it must be separated in the ground at least 20 feet from any other system or equipment grounding electrode.) See IEEE 142 for additional information on this grounding system. Voltage differences that may appear on the metallic frames of this equipment, due to the proximity of other system grounding connections, can be hazardous to the equipment operator.

The grounding electrodes identified in 250.52 include a metal underground water piping system which is in direct contact with the earth for 10 feet or more. Due to the wide use of nonmetallic water piping systems, the metal underground water pipe must be supplemented by at least one additional grounding electrode, which may be a ground rod. However, if a ground rod is used as the supplemental grounding electrode, the resistance to ground must not exceed 25 ohms (250.53(2)). If so, another electrode must supplement the single rod. The connection of the grounding electrode conductor to the water pipe must be made within 5 feet of the water pipe point of entrance into the building. This is so that a long portion of the interior metal water pipe is not used as an extension of the grounding electrode conductor. If this were done, and repairs to the interior metal water pipe were made, and metal pipe was replaced with nonmetallic pipe, the connection to ground would be lost. There is an exception here that applies to commercial, industrial, and institutional buildings, where the maintenance of these buildings is such that only qualified persons service the installation. In this case, the connection to the interior metal water pipe may be made beyond the first 5 foot point of entrance of the metal water pipe into the building. However, if this were done, the entire interior metal water pipe must be exposed, with an exception for a condition where the pipe is run through walls, floors, or ceilings.

The next grounding electrode includes one or more metal in-ground support structure(s) in direct contact with the earth, vertically, for 10 ft or more (3.0m). The metal support may, or may not, be encased in concrete. And it may consist of a structural steel column, as well as metal pilings or metal casings. This would be the most common means of using the metal support structure as a grounding electrode.

The next grounding electrode, which has been in the NEC since 1968, is a concrete-encased electrode. This consists of an electrode encased in 2 inches (5.08cm) of concrete, located within and near the bottom of a concrete footing or foundation, that is in direct contact with the earth, consisting of one or more bare or zinc-galvanized steel reinforcing bars or rods of not less than ½ inch (1.2cm) (#4 Rebar) in diameter, or consisting of at least 20 feet (6m) of bare copper conductor not smaller than No. 4 AWG (21.15mm²). The reinforcing bars or rods are permitted to be bonded together by the use of steel tie wires, welding, or the exothermic welding process. The reinforcing rods and the concrete become the grounding electrode, as well as the soil surrounding the concrete. One item of concern here is that the concrete footing will absorb and retain moisture from the soil surrounding it. In other words, the concrete is hygroscopic. In the event of a lightning strike to structural steel, this lightning current would flow through the steel (columns) and into the footing or foundation. The heat energy of this lightning current will cause the rapid expansion of the absorbed moisture in the footing or foundation and possibly cause significant damage. A means of protection from the hazards of ground-seeking currents would be to extend a copper conductor (No. 4 AWG (21.15 mm²) copper conductor) to a nearby ground rod or grid. This would provide another conducting path for the ground-seeking current to flow into the earth, instead of solely through the absorbed moisture in the concrete. This external connection to the rod would provide another conducting path into the earth for the lightning current to flow, and thereby provide some degree of protection for the concrete footing or foundation.

It should be noted that where the reinforcing rods in the footing or foundation are accessible (new building or structure), Section 250.50 requires their use as a part of the grounding electrode system, and the reinforcing steel would be bonded to any of the other available grounding electrodes. In addition, the connection between the copper grounding electrode conductor and the steel reinforcing rod must be listed for this purpose, or exothermically welded.

A section of rebar may extend from the footing or foundation, above grade, to accommodate the connection of the grounding electrode conductor.

And, it should be further noted that during tests that have been done, currents in the range of 500-2600 amperes have caused significant damage to the concrete. For this reason, the use of the supplemental external ground rods are very important.

Utilities have recognized this problem for many years. The concrete footings of transmission towers have been damaged due to the lack of proper bonding between the steel tower and the reinforcing rods of the footing.

Sometimes, this concrete-encased electrode is referred to as a 'Ufer Ground', in recognition of Mr. Herb Ufer. Mr Ufer did ground resistance testing of this concrete-encased electrode beginning in 1942. There were munitions facilities under construction at that time in Tucson, and Flagstaff, Arizona. And, Mr. Ufer tested each of these facilities, twice monthly, for

18 years (1942-1960). He gathered the data during this period and the average resistance-to-ground for all of these measurements was 3.57 ohms. For additional information in protecting this electrode from damage due to ground seeking currents, see IEEE F77-115-9. It is vital that the concrete footing or foundation is in direct contact with the earth. There must not be any separation caused by a vapor barrier.

The ground ring grounding electrode consists of a solid 2 AWG ($33mm^2$), or larger, bare copper conductor that completely encircles the building or structure at a depth of at least 30″ (750 mm).

It would be normal for the ground ring to be supplemented with driven ground rods in order to lower the resistance-to-ground because of possible poor soil conditions at the 30″ depth (750mm). The ground ring is commonly used for the grounding electrode of a lightning protection system (NFPA 780). The ground ring conductor should be solid and not stranded. This conductor will resist corrosion and have a longer service life, because it exposes less of its cross-sectional area than a stranded wire. Also, depending on soil conditions (corrosivity), it may be necessary to use a tinned copper conductor for corrosion protection. The use of a larger conductor for the ground ring is up to the designer or installer.

Now, we get to the rod and pipe electrodes that are so commonly used. These rod and pipe electrodes are to be at least 8 feet (2.4m) in contact with the soil and consist of the following materials.

   a. Electrodes of pipe or conduit must be at least 3/4″ in diameter, and, where of iron or steel, they are to have their outer surface galvanized or otherwise protected against corrosion.
   b. Electrodes of rods of iron or steel are to be at least 5/8″ in diameter. Stainless steel rods less than 5/8″ in diameter, nonferrous rods, or their equivalent shall be listed (UL 467), and shall be at least 1/2″ in diameter.

The resistance-to-ground of these rods or pipes are the sum of three components.

**First** - the resistance of the rod or pipe electrode, which, of course, will be negligible.

**Second** - the resistance of the metal-soil interface, which is also negligible, unless some type of nonconductive coating is on the outside of the pipe or rod.

**Third** - the soil resistivity, which will be the major portion of the resistance-to-ground of the electrode. Therefore, we should be concerned about the soil resistivity, which may vary considerably from place to place. It would make sense to test the soil resistivity before the installation of the rod or pipe electrode.

Using a grounding megohmmeter, which is an AC device as opposed to DC in order to minimize the effects of stray DC currents in the earth due to galvanic action, we may determine the soil resistivity.

Four 1-foot long (30.48cm) metallic electrodes are installed in a straight line about 20 feet (6.0m) apart. The terminals on the meter (C1, P1, P2, C2) are connected to these short rods. The resistance is measured and the soil resistivity is calculated (Rho = $2\pi AR$ ).

A, in this formula, is the distance between the rods in centimeters (20 feet × 12 inches × 2.54 equals 610 centimeters), and (R) is the resistance recorded by the meter. For example, if the resistance recorded by this meter is 4 ohms, the soil resistivity would be 15,323 ohm centimeters − (6.28 (2 times 3.14) × 610 centimeters × 4 ohms). An ohm-centimeter is the resistance, in ohms, across the faces of a cubic centimeter of soil. According to IEEE-142, where the soil resistivity is 15,323 ohm centimeters, and a ground rod is driven into this soil, we will be able to calculate the approximate resistance-to-ground of the rod.

For example:

$$1/2" \,(1.27cm)\,diameter\,rod = \frac{15.323\,ohm - centimeters}{292} = 52.48\,ohms$$

$$5/8" \,(1.59cm)\,diameter\,rod = \frac{15.323\,ohm - centimeters}{302} = 50.74\,ohms$$

$$3/4" \,(1.90cm)\,diameter\,rod = \frac{15.323\,ohm - centimeters}{311} = 49.27\,ohms$$

Where these rods are 10 feet long (3m), and if we used an average divisor of '300', the resistance-to-ground of the rods will be about 51 ohms. Section 250.53(A)(2), Exception, would permit the resistance-to-ground of a single rod, pipe, or plate to be as high as 25 ohms, without being supplemented with an additional grounding electrode. In order to reduce this resistance-to-ground, if may be necessary to install parallel grounding electrodes (rods). If this is to be done, the spacing between these parallel rods needs to be considered. For example, if a rod is driven into the earth to a 10 foot depth (3 meters), the sphere of influence of the rod extends approximately 10 feet radially around the rod, and 10 feet (3 meters) beneath the rod. Therefore, if two rods were driven 10 feet deep into the earth, these rods should be 20 feet apart (6 meters), so that one rod does not enter the sphere of influence of the second rod. In so doing, their combined resistance-to-ground is about

1/2 of the single rod resistance-to-ground. Closer spacings will mean higher resistance-to-ground. The 6 foot spacing (1.5 meters) between electrodes referenced in 250.53(A)(3) is not enough. See 250.53(A)(3) Informational Note)

Although the spacing of 2 rod lengths between paralleled ground rods would normally yield a resistance-to-ground of approximately 50 percent of a single rod, maintaining this spacing for additional ground rods may not always be possible due to property (or other) restrictions.

A possible solution may be to drive the rods to a greater depth, where the soil may be of a lower resistivity due to its composition, and other factors, including temperature and moisture.

With the increased depth and lower soil resistivity, it may be possible to reduce the spacing between paralleled rods to less than 2 rod lengths.

Driving rods deeper into the soil may reduce the resistance-to-ground, if this is possible. As a good rule-of-thumb, if you double the depth of the rod, you will reduce the resistance-to-ground by about 40 percent. Depending on soil conditions, increasing the depth of the rod to reduce the resistance-to-ground of the electrode, is very beneficial.

When using a point-to-point method of checking the resistance-to-ground of an existing grounding system, it is necessary to disconnect this system from the electrical installation. So, an alternate grounding electrode (system) must be used while the test is being conducted.

Fortunately, there are clamp-on meters available that permit the resistance-to-ground test to be done without the removal of the grounding electrode conductor. However, this type of tester may not be used for a single ground rod, as a 'loop' must be established for the reading to be accurate.

Soil enhancements will serve to lower the resistance-to-ground of the rod. For example, Erico 'GEM' is a material, in package form, that has a very low resistivity, and can be used to treat the soil, whether the ground rod is driven vertically, or laid horizontally in a trench. For more information, contact Erico at 1-800-248-WELD, and also ask them for information relating to grounding AC substations and signal-reference grids for Information Technology Equipment rooms (Article 645).

As we stated earlier, it is the soil resistivity that determines the resistance-to-ground of the ground rod. I would like to use an example of a typical ground rod, which is driven 10 feet, (30.48 meters) vertically, into the earth. There are concentric earth rings that surround this rod, extending away from the rod.

The first concentric earth 'shell' is .1 foot (3.048cm) away from the rod. This earth 'shell' represents 25 percent of the total resistance-to-ground of the rod.

The next earth 'shell', which is .2 foot (6.096cm) away from the rod, represents 38 percent of the total resistance-to-ground of the rod. We can see here that the soil immediately surrounding the rod has a lot to do with its resistance-to-ground.

The next earth 'shell', which is .5 foot (15.24cm) away from the rod, represents 52 percent of the total resistance-to-ground of the rod.

The next earth 'shell', which is 1 foot (30.48cm) away from the rod, represents 68 percent of the total resistance-to-ground of the rod.

The next earth 'shell', which is 5 feet (152.4cm) away from the rod, represents 86 percent of the total resistance-to-ground of the rod.

The next earth 'shell', 10 feet (304.8cm) away from the rod, and the rod is 10 feet deep, represents 94 percent of the total resistance-to-ground of the rod. Therefore, we can say that the depth of the rod establishes its earth 'shell', and no other grounding electrodes should occupy this 'shell'. This is why we say that the sphere of influence of the rod is established by its depth.

There are chemically-charged rods (possibly, magnesium sulfate) available (UL 467J). These are hollow core rods that are filled with a low resistivity material, with holes drilled around the periphery of the rod. These rods have an outside diameter of at least 2-1/8" (5.40cm). Holes would be drilled into the earth and the rods are surrounded with a backfill material that is supplied by the manufacturer, which has a very low resistivity. The material within the rod reacts with the normal moisture in the air, dissolves and slowly migrates through the soil, adding crystal root structures in the soil, and lowering the soil resistivity. These rods normally have a service life of up to 50 years, and require no maintenance.

The diameter of the rod has little to do with its resistance-to-ground. This is due to the fact that there would not be much more surface contact area from a ½ inch diameter rod than you would get with a 1 inch diameter rod. By doubling the diameter of the rod, the cost and weight of the rod is increased by 400 percent, and the resistance-to-ground of the rod is decreased by about 9.5 percent. Certainly, the diameter of the rod has a lot do with its tensile strength, but very little to do with its resistance-to-ground.

Lastly, the concentric earth rings that surround the rod have everything to do with the resistance-to-ground. Therefore, make sure that the rod is installed in a way that these earth rings are established. This means that the rod is not too close to the building foundation, where these earth rings are not established on one side of the rod.

If we can imagine the same 10 foot long rod in our example, and we assume the same percentages associated with the concentric earth shells surrounding the rod, and a lightning current of 15,000 amperes is carried into the 10 foot deep rod (3.048m), and the resistance-to-ground of the rod is 20 ohms, the instantaneous voltage-rise on connected systems and equipment would be 300,000 volts (15,000 amperes × 20 ohms) above earth potential. Therefore, 25% of this voltage would be dropped through the first earth shell, 38% through the second earth shell, 52% through the next earth shell, 68% through the next earth shell, 86% through the next earth shell, 94% through the next earth shell, etc. If someone were standing close to this rod, and this lightning current was flowing into the earth, due to these percentages, the voltage difference between their two separated feet would be on the order of tens of thousands of volts (step potential). This is the reason for the low-resistance-to-ground system in a typical electrical substation. In these installations, a ground grid is installed in order to create an equipotential plane so that step and touch potential differences are limited, instead of connecting systems and equipment to ground independently. Any metal fence that surrounds the substation would be connected to this ground mat at a maximum of 160 feet (50 meters), as well as at the corner posts. Any discontinuous sections of the fence, such as at gates, would be bonded across to reduce voltage differences, and bonded to the ground mat. Every time a connection is made to this ground grid, a ground rod would be installed. This would include system and equipment grounding connections. The metal fence would be bonded to the ground grid on both sides where overhead lines pass over the fence. It would be improper to extend this metal fence to a remote area, such as a parking lot, because any voltage-rise through the fence would be extended through the fence to this remote area. Therefore, it would be important to install a nonmetallic barrier, such as a section of wood fencing, or another effective means to interrupt the electrical continuity of the fence, in order to eliminate this voltage-rise (NEC Section 250.194) (IEEE STD. 80).

The average service life of a typical copper-clad steel rod is 30 years. UL has done extensive testing of rods in different types of soil. In 41 of 43 different types of soil that were tested, the corrosion penetration after 30 years was .17mm. The copper sheath of these rods is .25mm thick to assure at least a 30-year service life. Ground rods are available that have a copper sheath that is .33mm thick, and an average service life of 40 years.

The next ground electrode is a buried metal plate, which may be ferrous or nonferrous. If it is of nonferrous material, the thickness of the plate is .06 inch (1.5mm). The plate would be at least one square foot so that it exposes at least 2 square feet (1.86m$^2$) to the soil (both sides). If it is of ferrous material (iron or steel), the thickness of the plate is to be at least ¼ inch (6.4mm). This plate is to be buried at least 2-1/2 feet (750mm) deep in the earth to satisfy 250.53(H). This depth may or may not be enough to provide a low resistance-to-ground connection, depending on soil conditions. However, the plate electrode does provide a better means of capacitive coupling to the earth than a ground rod, due to its flat surface and greater cross-sectional area. This is beneficial for a lightning protection system.

Also, Section 250.52(A)(8) permits other local underground systems or structures, such as metal piping systems and underground tanks, as well as a metal well casing (that is not bonded to a metal water pipe) to be used as grounding electrodes.

Please refer to 250.53 for information regarding grounding electrode system installation. Also, see 547.9 and 547.10 for special bonding and grounding requirements for Agricultural Buildings.

Section 250.54 permits a supplemental (auxiliary) grounding electrode (there may be more than one) to be connected to the equipment grounding conductors (250.118). The supplemental grounding electrode(s) is not subject to the bonding jumper size of 250.53(C), or to the resistance-to-ground requirements of 250.53(A)(2),Exception. Of course, this supplemental grounding electrode(s) does not take the place of the equipment grounding conductor that is run with the circuit conductors (250.134(B), (300.3(B)). The lowest impedance path for the equipment grounding conductor is when the equipment grounding conductor encloses, or is run with the circuit conductors, and is in close proximity to the conductors of the same circuit. Remember, magnetic-flux density is inversely proportional to the square of the distance between the adjacent conductors. As the spacing between conductors (AC) of the same circuit increases, the magnetic field strength decreases, and this results in increased circuit impedance.

I remember when I was working on a large construction project many years ago, and the electricians were preparing to extend branch circuits in nonmetallic conduits for the parking lot lighting (metal poles) standards. Each metal pole was to be connected to a ground rod for lightning protection. And it was believed by these people that there was no need to install equipment grounding conductors in these nonmetallic conduits, due to the installation of the ground rods. If this had been done, the only path for ground-fault current to flow would be through the earth between the ground rods in the parking lot and the grounding electrodes(s) at the transformer that supplied these circuits. Clearly, this would have been a high impedance path, and if a ground-fault ever developed, the fault current would not be high enough to clear even small overcurrent devices. Fortunately, this condition was recognized before the installation of the branch circuits.

Section 250.64(B) discusses the securing and protection of grounding electrode conductors. Normally, a No. 8 AWG (8.37mm$^2$) grounding electrode conductor requires physical protection. A No. 6 AWG (13.3mm$^2$) grounding electrode conductor that is free from physical damage requires no physical protection, and this applies to grounding electrode conductors that are No. 4 AWG (21.15mm$^2$), and larger. If the physical protection is in the form of a ferrous metal raceway, it would be imperative to bond the raceway on both ends to the grounding electrode conductor, so as not to create a 'choking' effect. If this were not done, the metallic enclosure (steel) would be a problem. As current flows through the grounding electrode conductor, due to fault conditions (or even normal conditions), the magnetic field surrounding the grounding electrode conductor would induce a current in the metallic enclosure. Unfortunately, this current is in a direction opposite of the original influence (inductive reactance). This is the 'choking' effect, and this condition would greatly increase the impedance of the grounding electrode conductor to the grounding electrode(s). By bonding the metallic enclosure to the grounding electrode conductor, in effect, the metallic enclosure would become the grounding electrode conductor, as much more current would flow through the enclosure. Please refer to 250.64(B),(E) (IEEE Paper 54).

Section 250.66 deals with the size of the grounding electrode conductor for AC systems. If the connection is to a rod, pipe, or plate, the conductor may be as small as No. 6 AWG copper (13.30 mm$^2$). If it is made to a concrete-encased electrode, that portion of the conductor that is the sole connection to the grounding electrode may be as small as No. 4 AWG copper (21.15mm$^2$). If the connection is to be made to a ground ring, that portion of the conductor that is the sole connection to the electrode may be as small as No. 2 AWG copper (33.63mm$^2$).

It should be noted, for example, if the grounding electrode conductor extends to a series of grounding electrodes, such as a ground rod, a concrete-encased electrode, a ground ring, and a metal water pipe, and the grounding electrode conductor for the water pipe, from Table 250.66, was required to be 1/0 copper (53.51mm$^2$), and this conductor extended to the ground rod, and then to the concrete-encased electrode, the ground ring, and finally to the metal water pipe, the GEC would be 1/0 copper, and not No. 6, No. 4, and No.2 (250.66(A),(B),(C)), (250.64(F)).

If the connection of the grounding electrode conductor is to be made to a metal underground water pipe (250.52(A)(1), or effectively grounded building or structural steel (250.52(A)(2), the grounding electrode conductor is to be sized from Table 250.66. I have always maintained that where the length of the grounding electrode conductor exceeds 100 feet (30m), the conductor size should be increased. We are relying on this conductor to hold our systems and equipment at, or near, earth potential. The longer this conductor is run, the less effective it is in performing this function. For example, let's say we are dealing with a service arrangement, or possibly a separately-derived system, and the size of the service conductors or feeder conductors are 500 kcmil copper. The grounding electrode for these systems is a metal in-ground support structure(s). If a copper grounding electrode conductor is to be used, Table 250.66 would specify a 1/0 AWG conductor size (53.51mm$^2$). But, if the grounding electrode conductor is to be run for 140 feet from the service equipment, or from the separately-derived system, the grounding electrode conductor size should be increased. If we make a reference to Chapter 9, Table 8-DC Resistance at 75°C for a 1000 foot length (304.8 meters), the resistance for a 1/0 uncoated AWG copper (53.51mm$^2$) conductor is 0.122 ohms. We would then reduce the DC resistance from 1000 feet to 100 feet, or .0122 ohms. If we take this 100 foot D.C. resistance and divide this value by the percentage of the grounding electrode conductor length, as compared to a 1000 foot length, we would then have the 1000 foot D.C. resistance of the grounding electrode conductor. And then we can determine the conductor size from the 1000 foot DC resistance values that are expressed in Table 8 of Chapter 9.

$$\frac{.0122\,ohms}{.14} = .0871 Ohms$$

In this case, the grounding electrode conductor size would be 3/0 AWG (85 mm$^2$) uncoated copper, due to its 140 foot length. This conductor has a 1000 foot DC resistance of .0766 ohms. You may also use this formula when you size equipment grounding conductors that are over 100 foot long (Table 250.122). For example, let's say that we have a 20 amp circuit and we are using uncoated copper stranded wire. Let's also say that the length of this circuit is 150 feet, from the source to the load. Typically, Table 250.122 requires a minimum equipment grounding conductor size of No. 12 AWG copper (3.30mm$^2$) for this 20 ampere circuit. However, due to the 150 foot length of this circuit, we would want to increase the conductor size to compensate for voltage-drop during ground-fault conditions (250.122(B)). The greater the voltage-drop in this equipment grounding conductor, the greater the voltage-rise on equipment frames during ground-fault conditions. Remember, 250.4(A)(5) requires you to limit this voltage-rise in order to provide an effective ground-fault current path. If we are installing a circuit using uncoated copper stranded conductors, Table 8 of Chapter 9 specifies that the 1000 foot DC resistance of an uncoated copper conductor (No. 12 AWG) is 1.98 ohms. We would then reduce this 1000 foot DC resistance to

100 feet or .198 ohms. Then, we would divide this 100 foot DC resistance (.198 ohms) by the percentage of the equipment grounding conductor length, as compared to a 1000 foot length (150 feet as compared to a 1000 foot length). The result will be the 1000 foot DC resistance of the conductor that we should use for the equipment grounding conductor.

$$\frac{.198\,ohms}{.15} = 1.32\,ohms\;(1000\,foot\,length)$$

In reviewing Table 8 of Chapter-9, the equipment grounding conductor would have a 1000 foot DC resistance of not more than 1.32 ohms, and referring to Table 8 of Chapter 9, we find the conductor size to be No. 10 AWG copper, which has a 1000 foot DC resistance (uncoated copper) of either 1.21 ohms, if it is a solid conductor, or 1.24 ohms if it is a stranded conductor. Of course, the other conductors of this circuit would be increased in size to reduce voltage-drop as well (210.19(A), Informational Note 4), (215.2(A)(1)(b), Informational Note 2).

If we base this calculation on the AC impedance values of Chapter 9, Table 9 for uncoated copper wire (Z at 0.85 PF), instead of the DC resistance values of Table 8, the 1000 foot value (as if it were installed in a nonmetallic conduit) for the 10 AWG wire is 1.1 ohms, as opposed to the DC resistance at 75°C. for the same 1000 foot length, which is 1.21 ohms (solid) or 1.24 ohms (stranded).

Now, let's compare the size of the copper equipment grounding conductor from Table 9. The 1000 foot AC impedance for No. 12 uncoated copper, as if it were within a nonmetallic conduit is 1.7 ohms. Reducing this to 100 feet, the AC impedance is .17 ohms, as compared to the 100 foot DC resistance of .198 ohms.

The circuit length is 150 feet. And this length is 15% of 1000 feet.

$$So,\;\frac{.17\,ohms}{.15} = 1.13\,ohms\;(1000\,feet)$$

Once again, checking the AC impedance values of Table 9, we find that the equipment grounding conductor is No. 10 AWG, uncoated copper, which has a 1000 foot AC impedance of 1.10 ohms.

We can see, that in this example, the physical size of the equipment grounding conductor is the same, whether Table 8 or Table 9 is used.

Now, we will make a comparison of this example by using the AC impedance values of Chapter 9, Table 9 to determine whether the grounding electrode conductor size will change based on the same conditions.

Once again, the ungrounded service or feeder conductors are 500 kcmil copper (253.35mm$^2$). The grounding electrode conductor from Table 250.66 is 1/0 AWG copper (53.51mm$^2$). And this conductor is to be extended to a length of 140 feet (42.67 meters) to the grounding electrode, which is a metal underground water pipe, or a metal in-ground support structure(s) (250.52(A) (1)(2)).

From Table 9, we select a value of 0.13 ohms for a 1000 foot length of 1/0 AWG uncoated copper. The 1000 foot impedance of 0.13 ohms is reduced to the impedance value for a 100 foot length (30.48 meters), and this equals .013 ohms. Then, we divide .013 ohms by .14, which represents the length of this conductor, as compared to 1000 feet (304.8 meters), (140 as compared to 1000).

$$\frac{.013\,ohms}{.14} .0928\,ohms$$

This value (.0928 ohms) will be the appropriate 1000 foot AC impedance of the grounding electrode conductor. A 3/0 AWG uncoated copper conductor (85mm$^2$) has a 1000 foot AC impedance of .088 ohms (PVC conduit).

So, the 3/0 AWG copper conductor (85mm$^2$) should be used.

In effect, the 3/0 AWG copper conductor at a length of 140 feet would be as effective as a 1/0 AWG copper conductor at a length of 100 feet, which is the basis for the selected wire sizes in Table 250.66.

I cannot emphasize enough that the basis of Table 250.66 and Table 250.122 is a length of 100 feet (30.48 meters), and the acceptable voltage-rise above earth potential that the 100 foot length represents.

By increasing the conductor size for a greater length, we will keep the intended purpose of these Tables intact.

## Review

Ungrounded service conductor size - 500 kcmil copper (253.35 mm$^2$)
Grounding Electrode Conductor Size – 1/0 AWG copper (53.51 mm$^2$)
Grounding Electrode (Metal In-Ground Support Structure(s) (250.52(A)(2))
(Underground Metal Water Pipe (250.52(A)(1))
Length of Grounding Electrode Conductor 140 feet (42.67 meters)
Grounding Electrode Conductor size from Chapter 9-Table 8 -
(DC resistance at 75°C.) - at 140 foot length - 3/0 AWG - Copper
Table 250.66-1/0 uncoated copper .122 ohms - 1000 feet
3/0 uncoated copper .0766 ohms - 1000 feet

Based on our calculation, the Grounding Electrode Conductor for the 140 foot length would require a DC resistance for a 1000 foot length of .0871 ohms. 3/0 AWG uncoated copper has a 1000 foot DC resistance (75°C.) of .0766 ohms. So, we increased the conductor size from 1/0 AWG uncoated copper to compensate for the additional 40 feet (12.19 meters) of length.

Using the AC impedance values of Chapter 9-Table 9 - the Grounding Electrode Conductor size is also 3/0 AWG uncoated copper (.088 ohms – 1000 foot length)

> 1/0 uncoated copper .13 ohms - 1000 feet
> 3/0 uncoated copper .088 ohms - 1000 feet

In addition, for the 20 ampere branch circuit, with a length of 150 feet (45.72 meters), a No. 10 AWG uncoated copper wire would be acceptable for the equipment grounding conductor, using Table 8 or Table 9.

Section 250.70 requires that the conductor connection of the bonding or grounding electrode conductor to the grounding electrode be made by a (listed) mechanical device, or one that is exothermically welded (250.8(A)). This connection is also required to be accessible (250.68(A)), unless it is encased in concrete, directly buried, or beneath fireproofing material on structural metal (250.68(A), Exceptions 1 and 2. And, as we have mentioned before, the connecting device must be listed for direct burial and for the connection of dissimilar metals.

For grounding and bonding requirements at services, we refer to 250.80. Normally, metal enclosures for service conductors are required to be grounded (250.80). However, where metal elbows are installed underground in an installation of rigid nonmetallic conduit, typically for wire pulling requirements, and the metal elbows are at least 18 inches deep (45.72cm), or the elbows are encased in 2" (5.08cm) of concrete to any part of the elbows, they do not require grounding (250.80, Exception).

Where underground service cable has a metallic outer covering that is metallically connected to the grounded-system conductor on the supply side of the service equipment, the outer sheath or armor is not required to be connected to the grounded system conductor at the building or structure. In this case the outer metallic covering may be insulated from any interior metal raceway or piping (250.84(A),(B)).

Normally, metal enclosures for other than service conductors require grounding. But, once again, if metal elbows are installed in underground installations that are nonmetallic, the elbows do not require grounding if provided with an earth cover of at least 18 inches (45.72cm) to any part of the metallic elbows, or the elbows are encased in at least 2 inches (5.08cm) of concrete (250.86, Exception No. 3). This would include underground feeders or branch circuits.

Another Exception (250.86, Exception No. 2) would permit short sections of metal enclosures used as physical protection and, or, support of cable assemblies and incased in not less than 2 inches(50mm) of concrete to not be grounded. Also, see 300.10, Exception No. 1. It would be up to the local authority to determine what a short section(s) means in relation to length.

In dealing with services, Section 250.92 requires that service raceways, cable trays, cablebus framework, auxiliary gutters, and cable armor be bonded together into a continuous electric conductor, except the cable armor or metal sheath as permitted by 250.84. Bonding of metal raceways used as service equipment is important because the service conductors are normally not provided with overcurrent protection on the line-side of the service disconnect. And, if there were a fault (short-circuit or ground-fault), this fault may have to cause the fusing or melting of the service conductors before the fault is cleared. Section 250.92(B) would not permit double locknut construction as a bonding means around concentric or eccentric knockouts in service equipment enclosures. A bonding bushing may be used around the double locknut construction to assure bonding integrity where these concentric or eccentric knockouts are encountered. The bonding jumper for the bonding bushing is a Supply-Side Bonding Jumper, and it is sized in accordance with 250.102(C)(1). Bonding-type locknuts may be used on clean knockouts in service equipment.

An accessible means external to the service equipment enclosure is to be provided in order to allow the bonding of other systems, such as the grounding system of communication systems, TV or radio equipment grounding systems, or the grounding electrode of a CATV system (250.94(A)).

Section 250.96(A) requires all metal raceways, cable trays, cable armor or sheath, metal enclosures, and all metal frames and fittings that are to serve as equipment grounding conductors, with or without supplementary equipment grounding conductors (such as when a metal raceway is supplemented with an internal equipment grounding conductor), to be bonded together to form a grounding circuit which is electrically continuous and has the capacity to safely conduct any ground-fault current that may be imposed. Once again, the ground-fault current must be known, as well as the operating characteristics of the overcurrent device (how fast it will clear the ground-fault), and the short-time current (insulation) withstand rating of the equipment grounding conductor).

Where required for the reduction of electrical noise (electromagnetic interference) on the grounding circuit, an equipment enclosure supplied by a branch circuit(s) may be isolated from a metal raceway containing circuits supplying only that equipment through the use of one or more listed nonmetallic raceway fittings, which are located at the point of attachment of the raceway to the equipment (250.96(B)). This isolated metal raceway is to be supplemented by an insulated equipment grounding conductor of the proper size (based on the rating of the largest overcurrent device protecting any circuit in the raceway). The equipment grounding conductors, such as the isolated metal raceway and the insulated equipment grounding conductor inside the metal raceway, are bonded together at the source of these circuits, but not at the equipment. This is the same provision that is applied to cord-and-plug connected equipment, which is connected to ground through the use of isolated grounding-type receptacles (250.146(D)). In these cases, the equipment grounding conductors, bonded at the source of these circuits, are purposely isolated at the equipment. Please check with local authorities before allowing this method of isolation, because there may be a significant voltage difference between the isolated metal raceway and the equipment frame, if a ground-fault occurs to the equipment frame.

Section 250.97 covers the equipment grounding and bonding of metal enclosures that contain other than service conductors where the voltage exceeds 250 volts-to-ground (480/277 volts) - (600/347 volts). These bonding and grounding provisions are the same as for services in 250.92(B), except that where oversized concentric or eccentric knockouts are not encountered, threadless couplings and connectors for cables with metal sheaths, or two locknuts on rigid or intermediate metal conduit, one inside and one outside of boxes and cabinets, as well as fittings with shoulders that seat firmly against the box or cabinet, such as electrical metallic tubing connectors and flexible metal conduit connectors, are an acceptable bonding means. If the box or cabinet has oversized concentric or eccentric knockouts and it is 'listed' for use without additional bonding means, the double locknut construction is acceptable as the bonding means.

In Hazardous (Classified) Locations, bonding is extremely important because loose connections in raceway systems may lead to arcing conditions during ground-faults. Of course, these arcing faults may prove to be a source of ignition in these special environments. All metallic conduit connections are to be made wrenchtight in Class I and Class II Hazardous (Classified) Locations. Listed threadless fittings for rigid metal conduit and intermediate metal conduit are permitted in Class I, Division 2 locations (501.10(B)(1)). Threadless fittings are acceptable in Class II, Division 2 locations (502.10(B) (1)), as well as Class III, Division 1 and 2 locations (503.10(A),(B)).

Also, see 505.15(A),(B),(C) for the wiring methods in Class I, Zone 0, Zone 1, and Zone 2.

Equipment bonding jumpers on the supply side of the service equipment are not to be smaller than specified in Table 250.102(C)(1), except where the service conductors are larger than 1100 kcmil copper (557mm$^2$) or 1750 kcmil aluminum (887mm$^2$). Then, the bonding jumper is based on 12½ percent of the area of the largest phase conductor. For example, where the service conductors are installed in parallel, let's say 2000 kcmil copper (1013mm$^2$) per phase, the bonding jumper would be at least 250 kcmil copper (127mm$^2$). Where these parallel conductors are installed in separate raceways, the equipment bonding jumper shall be sized in accordance with the size of the ungrounded service conductors in each raceway, as referenced in Table (250.102(C)(1)).

Where a metal raceway or cable armor surrounds a grounding electrode conductor for physical protection, as we have discussed earlier in this Chapter (250.64(E)), the bonding jumper on each end of the metal enclosure must be at least the size of the grounding electrode conductor (IEEE Paper No. 54).

The equipment bonding jumper on the load side of the service overcurrent device is to be sized, at a minimum, in accordance with 250.122 (250.102(D)).

Equipment bonding jumpers may be installed on the inside or outside of a raceway or enclosure. If installed on the outside, their length is limited to 6 feet (1.8m), and the bonding jumper is to be routed with the enclosure for important impedance reduction. Where the bonding jumper is installed inside of a raceway, the provisions of 250.119 and 250.148 shall apply. This means that the bonding jumper is properly identified if it is insulated or covered. The identification must have a continuous outer covering that is green, or green with one or more yellow stripes. If the conductors are larger than No. 6 AWG (13.30 mm$^2$), the insulation or covering may be identified at each end, and where the conductor is accessible, such as in a junction box. This identification must surround the conductor and the exposed insulation or the covering is to be green in color, or the exposed insulation or covering may be marked with green tape or adhesive labels, or stripping the insulation or covering from the entire exposed length. This same means of identification would apply to multiconductor cables, regardless of conductor size, where qualified people service the installation and, where green tape or green adhesive labels are used, the tape or label must encircle the conductor (250.119(A),(B)).

Where equipment grounding conductors are spliced inside a box or terminated on equipment within or supported by a box, proper devices for splicing and terminating shall be used (110.14(B)). The arrangement of the grounding connections is to be such that the removal of a receptacle or device will not interrupt the continuity of the grounding conductors (250.148(B)).

Of course, this provision does not apply to the isolated grounding-type receptacles referenced in 250.146(D). A similar provision relating to the continuity of grounded (neutral) conductors in multiwire branch circuits appears in 300.13(B). And don't forget that sheet metal screws may not be used to secure equipment grounding conductors to enclosures (250.8(A)).

The various types of equipment grounding conductors are referenced in 250.118. This is an extensive section, so take several minutes to review it. Also, please refer to 517.13(B) where it is required to provide an insulated copper conductor inside metal raceways or metal-clad cables for the grounding terminals of receptacles and all noncurrent carrying conductive surfaces of fixed electrical equipment likely to become energized, and where subject to personal contact and operating at over 100 volts, in patient care spaces in a health care facility. The metal-clad (Type MC) cable with the interlocking flexible metal outer covering has an insulated grounding conductor inside the cable assembly, due to the fact that the metal outer covering of this cable is not listed as an equipment grounding conductor (330.108). However, there is an MC cable that is listed as an equipment grounding conductor. This cable has an aluminum bonding strip beneath the metal interlocking armor, similar to Type AC cable.

Section 250.122 deals with the overall size of equipment grounding conductors. And Table 250.122 identifies the <u>minimum size</u> of equipment grounding conductors that are copper, aluminum, or copper-clad aluminum. Of course, the size of the equipment grounding conductor may have to be increased due to circuit conditions, such as where an increase in the size of the ungrounded conductor(s) takes place because of voltage-drop, or high ambient temperatures, or proximity effects (the number of current-carrying conductors in a raceway or cable). This increase in size may be due to an increase in the size of the circuit.

Section 250.122(B) requires a proportional increase in the size of the equipment grounding conductor where a change occurs in the size of the ungrounded conductor(s).

However, a change in the 2020 NEC permits the equipment grounding conductor size to be determined by a qualified person. The authority having jurisdiction must determine the qualifications of the qualified person

For example, let's say that we have a 60 ampere circuit, which normally would require a No. 6 AWG copper conductor ($13.30 \text{ mm}^2$). However, due to voltage-drop, we are installing a No. 3 AWG copper conductor ($26.66 \text{ mm}^2$). Referring to Chapter 9, Table 8 'Conductor Properties' we find the circular-mil area of these conductors:

$$\frac{No.3\,AWG = 52.620\,cm}{No.6\,AWG \ = \ 26,240\,cm} = 2.0053$$

We would divide the circular-mil area of the No. 3 AWG conductor by the circular mil area of the No. 6 AWG conductor, as shown. Then referring to Table 250.122, we find that the minimum size of copper equipment grounding conductor for a 60 ampere circuit is No. 10 AWG. Once again, referring to Chapter 9, Table 8, we find that the circular-mil area of a No. 10 AWG conductor is 10,380 cm. We would then multiply 10,380 cm by 2.0053.

$$\begin{array}{r} 10,380 \ cm \\ \times 2.0053 \\ \hline 20,815 \ cm \end{array}$$

This is the cross-sectional area of the copper equipment grounding conductor that we would install. Again, by referring to Chapter 9, Table 8, we see that the equipment grounding conductor should be No. 6 AWG copper. Table 250.122 indicated that the minimum size copper equipment grounding conductor for a 60 ampere circuit is No.10 AWG ($5.26mm^2$). But due to voltage-drop, the equipment grounding conductor is No. 6 AWG copper, because of the 26,240cm area of this conductor.

For motor circuits, where the overcurrent device is an instantaneous-trip circuit breaker or motor short-circuit protector, which may be set at up to 8 times (or more) of the motor full-load current rating (430.52), the equipment grounding conductor size is permitted to be based on the minimum rating of a dual element time-delay fuse for the motor circuit (250.122(D)(2)).

Where conductors are installed in parallel, in multiple raceways or cables (310.10H), the equipment grounding conductors, used in nonmetallic raceways or cables, or where used to supplement metallic raceway systems, are also to be installed in parallel in each raceway or cable. Each parallel equipment grounding conductor is to be sized in accordance with the ampere rating of the overcurrent device protecting the circuit, with a reference to 250.122(F)(1)(b).

Where multiconductor cables are installed in parallel, within the same raceway, auxiliary gutter, or cable tray, a single equipment grounding conductor, based on the size of the circuit overcurrent device, may supplement the equipment grounding conductors within the multiconductor cable assemblies (250.122(F)(2)(b)).

Don't forget that the equipment grounding conductor that you provide has to establish an effective ground-fault current path as required by 250.4(A) (5),(B)(4).We discussed this earlier in this Chapter. It does not make any difference which type of equipment grounding conductor(s) we provide (250.118), this path must be permanent and continuous (300.10), (300.6(A), it must have ample capacity to carry any ground-fault current that may be imposed on it, and it must provide a sufficiently low impedance path that will limit the voltage-to-ground during ground-faults (the voltage-rise above earth potential ('0' volts), and also to facilitate the operation of the overcurrent device, if it is a solidly grounded system.

In DC circuits, where conductor resistance (and not inductive reactance) limits the flow of current through conductors, increasing the equipment grounding conductor size to compensate for voltage-drop is not a requirement. For example, a reference to 690.45 for PV Source and PV Output Circuits, addresses this provision.

Section 250.130 deals with the method of grounding equipment. Normally, this method of grounding equipment is accomplished by an equipment grounding conductor that is run with the circuit conductors, or encloses the circuit conductors, such as a metal raceway. Also, where steel or aluminum cable trays are installed, the use of the metal cable tray as an equipment grounding conductor is determined by 392.60. Where the metal cable tray is used for this purpose, proper bonding of the joints and sections is required (392.60)(B)(4)).

One problem that I have here is with Section 250.130(C)(1), where it would be permitted to install normal grounding-type receptacles, (available since 1953), as replacements for nongrounding-type receptacles, where an equipment grounding conductor is not provided in the branch circuit. In this case, the grounding conductor from the grounding-type receptacle grounding terminal may be connected to any accessible point on the grounding electrode system as described in 250.50. If this were done, the path for ground-fault current would be through the grounding electrode system, which is typically widely separated from the circuit conductors that supply the receptacle circuit. Because of this separation, the impedance of the equipment grounding circuit would have to be increased because of the loss of magnetic and capacitive coupling effects. Therefore, if a ground-fault were to occur in equipment that is connected to this receptacle circuit, the ground-fault current may be reduced to a level that would not be capable of interrupting even a small overcurrent device, and, would not provide an effective ground-fault current path required by 250.4(A)(5).

Of course, the use of a ground-fault circuit-interrupter type of receptacle may be used, because this device does not require an equipment grounding conductor to function, but functions due to the differential current between the ungrounded conductor and the grounded conductor. Please see Section 406.4(D)(2) for the use of the ground-fault protection of personnel requirements for these circuits that do not have an equipment grounding conductor. Also be aware that if this type of protection is provided, and a surge-protective device is installed into this receptacle, the SPD would not operate properly, because there is no equipment grounding conductor in the circuit.

If a nongrounding-type of receptacle(s) is replaced with a GFCI type of receptacle(s), the receptacles(s) is to be marked "No Equipment Ground". If nongrounding type receptacles are replaced with grounding-type receptacles that are supplied through a GFCI (receptacle or circuit breaker), these receptacles are to be marked "GFCI protected", as well as, "No Equipment Ground" (406.4(D)(3), Exception).

Section 250.134 addresses the means of grounding the metal frames of equipment that are fastened in place or connected by a permanent wiring method, as opposed to cord-and-plug connections, which are covered in 250.138. Section 250.134(A) recognizes any of the equipment' grounding conductors referenced in 250.118 for this equipment grounding purpose. 250.134(B) recognizes the use of an equipment grounding conductor contained within the same raceway or cable with the other circuit conductors. As we have discussed earlier for AC circuits, the impedance of the equipment grounding conductor is affected by its spacing from the other conductors of the same circuit. The lowest impedance path would be where the equipment grounding conductor is in close proximity to the other conductors of the same circuit. The magnetic-field strength between conductors is at its greatest level when the circuit conductors are close together. Therefore, it is imperative that this grounding conductor is inside the raceway or cable. You will recall that there is an exception to this condition in 250.102(E), where an external equipment bonding jumper may be on the outside of the raceway where its length is limited to 6 feet. This may be relatively common at motor locations where short sections of flexible raceways connect to the motor terminal housing. Also, 250.102(E)(2) has an Exception where external bonding connections may be longer than 6 feet, such as at outside pole locations for the purpose of bonding or grounding isolated sections of metal raceways or elbows used for pole risers.

We do not normally ground equipment frames with the grounded conductor, usually a neutral (remember that a Corner-Grounded Delta System has a grounded conductor that is not a neutral conductor). An Exception of Section 250.140 recognizes this as a possibility for the frames of ranges and clothes dryers, but only in existing installations. This concept dates back to World War II, where using the neutral conductor as the equipment ground for this equipment was recognized as means of saving copper. There are some restrictions here for these existing installations:

1. The circuit is 120/240 volt single-phase, 3-wire, or 208Y120 volt derived from a 3-phase, 4-wire, Wye connected system.
2. The grounded conductor of the branch circuit is at least No. 10 AWG copper (5.26mm$^2$) or No. 8 AWG aluminum (8.37mm$^2$).
3. The grounded conductor is insulated or bare and part of a Type SE service entrance cable and the branch circuit originates at the service equipment.
4. The ground contacts of receptacles furnished as part of the equipment are bonded to the equipment.

Section 250.142(A) also addresses conditions where the grounded conductor may be used as a means for grounding equipment.

1. On the supply side, or within the enclosure of the AC service disconnecting means.
2. On the supply side, or within the enclosure of the main disconnecting for separate buildings in accordance with 250.32(B). This is where two or more buildings or structures are supplied from a common service, which we covered earlier in this Chapter.
3. On the supply side or within the enclosure of the main disconnecting means or overcurrent devices of a separately-derived system in accordance with 250.30(A)(1). In this case, the connection of the grounded conductor to the metal frame of equipment would be done at the source of the separately derived system, or at the first disconnect or overcurrent device supplied from the separately derived system. And, from this connection, where the System Bonding Jumper is installed, the grounding electrode conductor would extend to the grounding electrode (system),

Section 250.146 covers the connection of a receptacle grounding terminal to a metal box. If the metal box is surface mounted, the contact between the metal box and the receptacle mounting strap is acceptable to ground the receptacle to the metal box. For cover mounted receptacles, this provision only applies to a box and cover combination that is 'listed' as providing adequate ground continuity between the cover and the metal box. These covers have been available for a number of years. Contact devices or receptacle straps that are 'listed' to provide adequate continuity between the receptacle mounting strap and the metal box are acceptable for flush mounted boxes. In this case, there would be no need to make a connection from the device grounding terminal and the metal box.

Section 250.162 covers the grounding of DC premises wiring systems.

Two-wire DC systems supplying premises wiring and operating at a voltage of 60 volts and not more than 300 volts are required to be grounded. There are exceptions here, for example, if the system is monitored by a ground detector, and supplies only industrial equipment in limited areas, it need not be grounded.

Also, a rectifier-derived DC system supplied from an AC system is not required to be grounded. Here, we have a connection with 250.20, which would permit the AC system supplying the rectifier to be ungrounded where the voltage range is from 50-1000 volts. Another exception is applicable to DC fire alarm circuits having a maximum current of 0.030 amperes, which may be ungrounded (Article 760, Part III).

The neutral of 3-wire, DC systems, supplying premises wiring is required to be grounded.

Where the DC system is supplied from a source of power which is onsite, a grounding connection shall be made at the source, or at the first disconnect or overcurrent device supplied from the source (250.164(B)).

The size of the grounding electrode conductor for a DC system, which is supplied by a 3-wire balancer set or winding, shall be at least the size of the neutral conductor, and, in any case, not smaller than No. 8 AWG copper or No. 6 AWG aluminum (250.166(A)).

When the DC system is not supplied through this balancer set or winding, the grounding electrode conductor is not smaller than the largest conductor supplied by the DC system, but at least No. 8 AWG copper or No. 6 AWG aluminum (250.166(B)).

The grounding electrode (system) may be any of the grounding electrodes referenced in 250.52 (250.166(C),(D), and (E)).

Where connected to a rod, pipe, or plate, the conductor that is the sole connection to the grounding electrode may be No. 6 AWG copper or No. 4 AWG aluminum (250.166(C)). However, 250.64(A) does not permit aluminum or copper-clad aluminum conductors to be terminated within 18 inches (450mm) of the earth.

A No. 4 AWG copper conductor, which is the sole connection to a concrete-encased electrode, is acceptable (250.166(D)).

If a ground ring is used as the grounding electrode, the grounding electrode conductor, which is the sole connection to the ground ring, shall be as large as the ground ring conductor, and at least No. 2 AWG copper (250.166(E) - 250.52(A)(4)).

For ungrounded DC separately-derived systems that are supplied from a stand-alone power source, possibly an engine-generator set, a grounding electrode conductor is to be connected to a grounding electrode in accordance with Part III of Article 250 (250.52–250.53). This connection to ground is to be made at the source of the separately-derived system, or at the first disconnect or overcurrent device supplied from the separately-derived system. And this is for the purpose of grounding metal enclosures, raceways, cables, and other noncurrent-carrying metal parts of equipment (250.169). Also, please refer to 250.34 for conditions where the metal frame or vehicle frame of a portable or vehicle-mounted generator may serve as the ground reference point, in lieu of making earth connections.

## Grounding of Systems or Circuits of over 1000 volts

A system neutral may be derived from a grounding transformer, typically a grounding zigzag autotransformer, which is used to ground the high voltage system (IEEE 142).

The minimum insulation level for neutrals of solidly grounded systems (no impedance device) shall be 600 volts where they are insulated. Of course, they may be bare (250.184(A)(1), Exception No. 1,2,3).

The neutral conductor ampacity must be suitable for the load, but no less than one-third of the phase conductor ampacity (250.184(A)(2)).

At these higher voltages, the solidly grounded neutral conductor is permitted to be grounded at more than one point. The connection to ground may be made at a transformer supplying conductors to a building or structure, as well as underground systems where the neutral is exposed, such as at subsurface enclosures, and where the conductors are installed as overhead circuits outdoors (250.184(C)).

At least one grounding electrode shall be connected to the multigrounded neutral conductor every 1300 feet, and the maximum distance between 2 adjacent electrodes shall be not more than 1300 feet (400 meters).

For impedance grounded neutral systems, in which an impedance device, usually a resistor (for a Wye connected system), limits the ground-fault current, the following applies (250.187):

1. Qualified persons service the installation.
2. Ground detectors are installed on the system.
3. Line-to-neutral loads are not supplied.

The grounding impedance is to be inserted in the grounding conductor between the grounding electrode and the neutral point of the transformer or generator.

The system neutral conductor is to be insulated for the maximum neutral voltage, which will be 57.7% of the phase-to-phase voltage, or the phase-to-phase voltage divided by 1.732.

Any equipment grounding conductors may be bare and connected to the ground bus and the grounding electrode conductor.

It is interesting to note here that the size of the neutral conductor is not specified, as it is for lower voltage systems in 250.36(B). Section 250.180 indicates that high voltage systems (over 1000 volts) must comply with all of the preceding Sections of Article 250 and with Sections 250.182 through 250.194. The neutral conductor from the neutral point of the transformer or generator should be sized in accordance with the current that may flow from this neutral point to the impedance device (resistor). It should be at least No. 8 AWG copper (8.37mm$^2$) or No. 6 AWG (13.30mm$^2$) aluminum, in order to comply with 250.36(B), but, of course, it may be larger than these conductor sizes if the ground-fault current is greater than these conductor ampacities.

Section 250.186 - Where an AC system operating at over 1000 volts is grounded at any point, a grounded conductor must be installed to each service disconnecting means and connected to the grounded conductor terminal or bus. A Main Bonding Jumper should connect the grounded conductor to the service disconnecting means enclosure (250.186(A)).

In this regard, Section 250.186(A) is the same as Section 250.24(C), where the AC system operates at 1000 volts or less.

This provision would provide a conducting path to the service point.

The grounded conductor is not to be smaller than required by Table 250.66 for the grounding electrode conductor, and not larger than the ungrounded conductors. Where the service conductors are larger than 1100 kcmil (557mm$^2$) copper or 1750 kcmil aluminum (887mm$^2$), the grounded conductor must not be smaller than 12½ percent of the cross-sectional area of the largest ungrounded conductor (250.186(A)(1)).

For a 3-phase, 3-wire Delta service, the grounded conductor (phase) is required to have the same ampacity as the ungrounded conductors (250.186 (A)(3)).

Section 250.187 applies to Impedance Grounded Neutral Systems for voltages over 1000 volts. These provisions parallel the requirements of Section 250.36 for systems of up to 1000 volts.

The grounding impedance, which is usually a resistor (but may be other devices, such as a zigzag autotransformer), is inserted in the grounding electrode conductor between the system neutral point and the grounding electrode (250.187(A)).

The neutral conductor will be insulated in accordance with the maximum neutral voltage, which is 57.7% of the phase-to-phase voltage (250.187(B)). This differs from Section 250.184(A)(1) for a solidly grounded neutral system where the insulation level for neutral conductors is permitted to be 600 volts. Of course, the impedance grounded system has no neutral conductor installed with the ungrounded conductors. But, there will be an equipment grounding conductor that is connected to the grounding electrode (system) (250.187(D)).

Portable or mobile equipment operating at over 1000 volts is supplied by an impedance grounded system where it is Wye connected, or through a neutral derived from a grounding (zigzag) autotransformer, if it is supplied from a Delta system (250.188(A)).

The grounding electrode that is used for the connection of the impedance device for the portable or mobile equipment must be separated in the ground by at least 20 feet (6.0m) from any other system grounding electrode (250.188(E), or equipment grounding conductor.

# 9

# Special Considerations

The normal grounding and bonding provisions for AC and DC systems that we have addressed in detail may not be effective in certain special applications.

For example, in dealing with the system grounding requirements for separately-derived power systems of Information Technology Equipment (NEC Article 645), the provisions of Parts I and II of Article 250 will apply. This includes Section 250.30, which covers the grounding of separately-derived AC systems. Section 645.14 states that 'power systems derived within listed information technology equipment that supply information technology systems through receptacles or cable assemblies supplied as part of this equipment, shall not be considered separately-derived for the purpose of applying 250.30'.

This provision has led to many unsafe installations where manufacturer's instructions have been misapplied or disregarded. And these separately-derived power systems, supplied by equipment, such as an isolation transformer, have been installed with no ground connection. Or, with a connection to a dedicated grounding electrode (system). This is certainly not acceptable.

It may be desirable to connect this system to an auxiliary grounding electrode (NEC 645.15-250.54), as opposed to grounded structural steel, possibly as a means of reducing common-mode noise in the grounding connection, that is, current from a grounded or neutral conductor to the equipment grounding system, due to the effects of capacitive coupling. Or, to reduce the possible effects of lightning-induced energy. But, if an auxiliary grounding electrode is used, this electrode must be bonded to the building grounding electrode system. The use of a dedicated and isolated grounding electrode (system) is not acceptable. The neutral point of the isolation transformer will be bonded to the metal frame of the transformer, which will be connected to the primary supply circuit equipment grounding conductor. So, the auxiliary grounding electrode will be common to the building grounding electrode system anyway.

In addition, it may be necessary to reduce the effects of higher-frequency signals (electromagnetic interference). Harmonic currents associated with the 3rd, 5th, 7th, 9th, 15th, etc, are especially of concern, as they may disrupt, or possibly destroy IT equipment. It may be necessary to install a Signal Reference Grid. This grid is constructed of flat, 26 gauge copper strips, which are 2 inches (.0508 meters) wide. The grid is typically arranged on 2-foot (0.6096 meters) centers. The grid may be 10, 12, or 16 feet (3.048; 3.6576, or 4.8768 meters) in width, and up to 100 feet (30.48 meters) in length. Other configurations may be manufactured as well.

The Signal Reference Grid is installed on top of the concrete subfloor and beneath the raised floor. All interconnecting cables, power-supply cords, and branch circuit wiring will on top of the SRG. Metallic raceways, metallic cable trays, metal air ducts, structural steel columns, and every sixth floor pedestal are bonded to the SRG. The metal frames of Power Distribution Units, electrical panelboards, transformers, and metal raceways, metallic cables, or metal cable trays that enter the Information Technology Equipment room are bonded to the SRG with short, flat, copper connections, which may be derived from the Signal Reference Grid.

And, finally, the SRG is bonded to the equipment grounding system that is provided for the Information Technology Equipment in order to create an equipotential grounding and bonding system within the room.

The skin effect of electrical conductors at high frequency, especially above 1 mHz, will cause round conductors to be ineffective. For example, a 4/0 AWG uncoated copper conductor (107.2 mm$^2$) has a DC resistance of 0.0608 ohms for 1000 feet at 75°C. And, an effective impedance of 0.074 ohms to neutral, per 1000 feet, at 0.85 Power Factor. However, a 4/0 AWG copper conductor has an effective impedance of 232 ohms for a length of 10 feet (3.048 meters) at 10 mHz.

Such a cable would not be able to equalize voltage across the ends of the cable.

In order to be effective as a bonding and grounding system, conductors must be flat and not round so as to expose as much of the conductor cross-sectional area to its surface as possible.

The Signal Reference Grid has proven to be an effective, low impedance protection system for sensitive electronic equipment. Its use as a supplement to a metallic raised floor system will also mitigate the effects of static electricity, which, in itself, may be damaging to sensitive equipment, as well as possibly causing data corruption.

Certainly, noise disturbances in the power and grounding system of sensitive electronic equipment can, and typically will, disrupt signals on interconnecting cables. Even though conductor resistance is much higher at frequencies above 1 mHz, due to skin effect, the more important concern is the inductive reactance of the complete grounding system at these higher frequencies. For this reason, the installation of a grounding electrode (system) with a low resistance-to-ground, possibly 2 ohms, or less, is not a magic cure-all for Communications or Information Technology Equipment sites. This is due to the fact that the grounding electrode conductor(s) that makes the connection to this low resistance-to-ground electrode may have such a high impedance that the low resistance-to-ground electrode may, in essence, be a moot point.

This means that as the frequency of the signal increases, the inductive reactance increases. This makes sense because of the fact that $XL = 2\pi FL$. This is apparent as the frequency reaches 100-200 Hz, and the inductive reactance of the circuit becomes more significant than the conductor resistance.

The frequency of lightning may range from 3 kHz to above 10 mHz. This is one of the reasons for the Informational Note of Section 250.4(A)(1), where a statement is made to the effect that a method of limiting imposed voltage is the routing of bonding and grounding electrode conductors so that they are not any longer than necessary to make the connection(s) and complete the installation so that unnecessary bends and loops are avoided. Increasing the length of these conductors and installing unnecessary, and especially sharp bends in these bonding and grounding electrode conductors will render them ineffective in reducing the voltage imposed by lightning, line surges caused by utility switching operations, or possible contact with higher voltage lines.

As we have already stated, as the signal frequency increases, and inductive reactance naturally increases, the overall impedance of the conductor that is used to make the earth connection may be extremely high, and, at that instant, a low resistance-to-ground grounding electrode system may not be effective in holding systems and equipment at earth potential.

This is not to say that a low-resistance-to-ground electrode (system) is not desirable. It certainly is, but additional protection techniques, such as, but not limited to, Surge Arresters and Surge-Protective Devices, are an important consideration.

Additional references should be noted including 280.12 (Surge Arresters), 285.12 (Surge-Protective Devices), 770.100(A)(4)(5), (Optical Fiber Cables and Raceways), 800.100(A)(4)(5), (Communications Systems), 810.21(E), (Radio and Television Equipment), 820.100(A)(4), (CATV and Radio Distribution Systems), and 830.100(A)(4)(5), (Network-Powered Broadband Communications Systems).

Each of these Sections mentions bonding and grounding electrode connections through conductors that are as short as practicable and run in as straight a line as practicable.

We know that noise transients are often caused by changes in load current, such as utility switching operations, the switching of large electrical loads, and, of course, the effects of lightning. A means of reducing the effects of transient voltage is through the use of Surge-Protective Devices, (UL 1449). This includes Type 1 at the service equipment, Type 2, possibly at the secondary side of a transformer (isolation) that supplies the IT system components, and Type 3 at the IT equipment location (285.23, 285.24, 285.25).

After considering this information, we can readily see that, due to higher signal frequency levels, and higher grounding conductor impedance, that is, both system and equipment grounding functions, it is difficult, if not impossible, to have a constant potential reference in this electrical environment. An excellent source of information on the subject of surge voltages in AC power circuits is ANSI/IEEE C62.41. This book documents the most severe transients on low-voltage AC systems.

So, we are primarily concerned with the electrical transient voltages from external sources and their effects on the operation of sensitive electronic equipment. The installation of a manufactured Signal Reference Grid behaves in much the same way as a typical printed circuit board, where the circuit components are installed in close proximity of a 'ground' plane. Installing the fabricated SRG on the top of the concrete subfloor, with the interconnecting cables and other wiring methods laying on top of the SRG, mimics the concept of the printed circuit board. And, where the cables are installed in close proximity to each other, the effects of magnetic and capacitive coupling will be maximized.

The cross members of the SRG are exothermically welded, using a 'clamshell' type crucible. This connection method is used in the field to bond one section to another to cover the entire concrete subfloor. In some cases, the SRG is installed in the walls of the room and connected to the floor grid. This protection method provides an effective shield for the entire room. In effect, the SRG provides multiple low-impedance conducting paths which are connected to the single-point ground within the room.

Listed fittings are available for connections to the SRG.

The bonding conductors which connect the IT equipment frames, transformer frames, the frames of panelboards, intersystem bonding terminations, and structural steel will be short, flat conducting members. The use of short lengths of the flat copper grid is an effective method of accomplishing the connections to the SRG.

For additional information on Signal Reference Grids, contact Erico at 1-800-248-9353 or Erico.com.

In the early days of IT installations, it became apparent that the normal bonding and grounding concepts, which were fine for the electrical systems and equipment at that time, were not suitable for this sensitive electronic equipment. And so, new ideas prevailed, and dedicated grounding electrode systems started to appear. This seemed to solve the problem. But, the designers and installers overlooked the fact that these 'isolated' grounding electrode systems were really not isolated at all. This is because the earth is a conductor (albeit, relatively poor). And, also, the use of these dedicated and isolated systems pose a significant threat to personal safety, as well as the potential for equipment damage.

In addition, as a means to equalize voltage differences, decisions were made to establish an equipotential bonding system consisting of No. 4 AWG copper conductors (21.15 mm²) on 4 foot centers (1.2192 meters), arranged in a grid pattern. This bonding grid was designed to cover the concrete subfloor. Metallic wiring systems, structural steel columns, metal air ducts, the metallic raised floor, and the metal frames of equipment within the IT room, were bonded to this bonding grid. A connection was made between the bonding grid and the single-point ground established for the IT room.

This type of bonding and grounding system seemed to solve the problem of data corruption and equipment damage. However, this system was only effective at relatively low frequency levels. For signal frequencies above 1 mHz, this bonding system proved to be ineffective because the bonding conductors were round and their impedance at higher frequencies made these conductors appear as a virtual open circuit, even though their cross-sectional area (No.4 AWG-41,740 circular-mils) was fairly large.

Of course, the metallic raised floor of the IT room could be used as a Signal Reference Grid. And it would provide a low impedance conducting path, even at higher frequencies, as compared to a fabricated SRG. In this case, the steel members of the raised floor would provide multiple conducting paths for high frequency signals to be conducted to ground and away from sensitive equipment.

The use of the raised floor as a SRG certainly appears to be an inexpensive choice as compared to the manufactured SRG.

But, the problem is not the issue of steel vs. copper, because at higher frequency levels, steel and copper behave about the same. The major difference is the fact that the manufactured SRG is designed to lay on top of the concrete subfloor, and all of the wiring, including branch circuits, interconnecting cables, etc., are on top of the SRG, and therefore, in close proximity to it. Capacitive-coupling effects are a benefit for this reason, as compared to the metallic raised floor, which is above, and separated from the under-floor wiring. And, the reinforcing steel in the concrete subfloor is also capacitively coupled to the SRG. Circulating currents in these reinforcing members would be capacitively coupled into the SRG and conducted to ground, and away from sensitive circuits.

Another potential problem of the use of the steel raised floor for this purpose is the possible loosening of the raised floor joints and coupling members over time, as people and equipment move across the floor. These loose connections may increase the overall impedance of the grid, and these loose connections are a source of electromagnetic interference (EMI).

## Solar Photovoltaic Systems

The system grounding requirements for PV systems involve the following grounding arrangements (690.41):

1. Two-wire PV Arrays with one functional grounded conductor.

    An Array is defined in 690.2 as a mechanically integrated assembly of modules or panels with a support structure and foundation, tracker, and other components, as required, to form a DC or AC power producing unit. A module is a complete environmentally protected unit consisting of solar cells, optics, and other components, designed to generate DC power when exposed to sunlight. Photons are 'particles of light' that cause electrons to move in a semiconductor. A panel consists of a collection of modules mechanically fastened together, wired, and designed to provide a field-installable unit.

    A Functional Grounded PV System has an electrical reference to ground that is not solidly grounded. For example, consider an arrangement of PV modules that generate DC power at the appropriate voltage and current. This is the Photovoltaic Power Source Circuit, and it extends from the power source to a common connection point(s) of the DC system. When there are two or more PV Source Circuits, a Direct Current Combiner is used to combine these circuits and provide for one DC circuit output. The Photovoltaic Output Circuit extends from the PV source circuit to the Inverter or to the DC utilization equipment. The Inverter Input Circuit includes the conductors that are connected to the DC input of the inverter. The Inverter is the equipment that converts the DC input to an AC output. The Inverter Output Circuit includes the conductors that are connected to the AC output of an Inverter and to the AC connected load.

    A Functional Grounded PV system is one where there is no solidly grounded conductor on the DC side of the inverter and the Equipment Grounding Conductor from the Inverter AC output circuit(s) provides the ground connection for the Ground-Fault Protection and equipment grounding of the PV arrays. The EGC is connected to the grounding electrode (system) on the AC side of the Inverter.

2. Bipolar PV arrays, with a functional ground reference (center tap). A Bipolar PV array has 2 DC outputs, each having an opposite polarity to a common reference point or center tap, and this is where the ground (earth) connection is made.
3. PV arrays not isolated from the grounded inverter output circuit.
4. Those PV arrays that are ungrounded.
5. Solidly grounded PV arrays as permitted in 690.41(B), Exception. In this case, this Exception permits the ground-fault protection to be omitted for PV arrays with not more than 2 PV Source Circuits, and with all PV system DC circuits not on or in buildings.
6. PV systems that use other methods that accomplish equivalent system protection in accordance with 250.4(A), with equipment listed and identified for the use. Section 250.4(A) covers electrical systems that are grounded (solidly), as well as the bonding of electrical equipment, and providing an effective ground-fault current path, which for a solidly grounded system, will facilitate the operation of the circuit overcurrent device during a ground-fault, or initiate an alarm on a high-impedance grounded system.

Section 690.41(B) requires Ground-Fault Protection for DC PV arrays, with the exception of the PV arrays, with not more the 2 PV Source Circuits, where all PV system DC circuits are not on or in buildings and the system is solidly grounded. The Ground-Fault protection must detect ground-faults in the PV array DC conductors and equipment, and the faulted circuits must be automatically disconnected, or the inverter must automatically stop the current flow to the output circuits, as well as isolate the PV system DC circuits from the ground reference in a Functional Grounded System.

The Ground-Fault Protection System is not provided as a means of personnel protection. It is meant to provide protection against fire hazards.

The connection to ground for any current-carrying conductor is made by the Ground-Fault Protective Device for functional grounded systems. For solidly-grounded PV systems, the DC circuit grounding connection is made at any single point of the PV Output Circuit, that is, from the PV Source Circuit to the Inverter.

Several years ago, PV systems were solidly grounded, that is the DC Source Circuit negative conductor was connected to a grounding electrode (system).

The recommendation was to make this connection to ground at a point that was close to the DC source (modules) to afford better protection from lightning. This is true for any supply system, DC or AC. Also, keep in mind that the PV source is not a constant voltage source due to variations in the ambient temperature. As the temperature decreases, the supply voltage increases (Table 690.7(A)).

The grounding electrode (system) tends to stabilize the voltage. This is not to say that the earth connection would limit this voltage-rise.

The grounding electrode (system) would also be used as a means of grounding the exposed noncurrent carrying metal parts of PV module (support) frames, as well as the metal frames of electrical equipment and conductor enclosures.

The AC supply system from the inverter would also be connected to a grounding electrode (system). This would include the grounded conductor (neutral) and the equipment grounding system.

Of course, both systems, DC and AC, could be connected to the same grounding electrode (system), which would include the grounding electrodes referenced in 250.52(A) and 250.166. If the DC and AC systems are connected to separate grounding electrodes, they must be bonded to limit potential differences between the DC and AC systems (250.50, 250.58).

Also, if the grounding electrode is a single ground rod, pipe, or metal plate, with a resistance-to-ground of over 25 ohms, the grounding electrode must be supplemented by an additional electrode, which may be an additional rod, pipe, or plate (250.53(A)(2)).

However, in more recent times, a Functional Grounded System is the preferred method of system grounding. In this case, the DC system is not solidly grounded and the Equipment Grounding Conductor for the Inverter AC output, which is connected to the grounding electrode, (system), either at the Inverter or further downstream, provides the necessary ground reference for the DC Ground-Fault Protection and the equipment grounding of the PV supply system. This includes the metal support structures of the PV arrays and any metallic enclosures that are a part of the DC supply. The Ground-Fault Protection will be part of the Inverter.

In areas of significant thunderstorm activity, it may be beneficial to connect the PV array support structures to an auxiliary grounding electrode (system) as a means of lightning protection (250.52, 250.54). If used, it would not be necessary to bond the auxiliary grounding electrode, as the bonding is already afforded through the equipment grounding system. Or, if the building or structure supports a PV array, 690.47(A) requires that the array be connected to a grounding electrode system that is installed in accordance with Part III of Article 250, (250.52(A)). For example, if the building or structure has a grounding electrode system consisting of metal in-ground support structures (250.52(A)(2)), the PV array support system will be bonded to this grounding electrode. In this way, the array equipment grounding conductors will have a connection to the grounding electrode system.

## Equipment Grounding

Just as in other electrical systems, the exposed noncurrent carrying metal parts of the PV equipment, including the PV module supporting frames, metal enclosures, and metallic raceways and cable assemblies are required to be grounded. Equipment grounding conductors must be installed within the same raceway or cable assembly or otherwise run with the circuit conductors (690.43(C)).

In this respect, this requirement is similar to 300.3(B), even though these circuit conductors from the PV arrays to the Inverter are <u>Direct Current</u> and not subject to the inductive heating of Alternating Current circuits, and the increased circuit impedance of AC circuits.

The size of the equipment grounding conductors for PV Source and PV Output Circuits are based on 250.122. This means that the EGC is sized in accordance with the size of the circuit overcurrent device.

However, the circuits may originate from the PV modules, where there are no overcurrent devices protecting these circuits. In this case, an assumed overcurrent device, in accordance with 690.9(B), will determine the size of the equipment grounding conductor(s).

Section 690.9(B) requires that the overcurrent devices that are used in <u>PV systems must be listed.</u> This means that these devices must be listed and identified for DC systems, which excludes overcurrent devices that are listed only for AC systems. Alternating current passes through zero two times during each cycle. Direct current is constant, and, especially during short-circuit or ground-fault conditions, these faults are more difficult to interrupt than in alternating current circuits.

For PV systems, the overcurrent devices must be rated at not less than 125% of the maximum currents calculated in accordance with 690.8(A).

Or, if an assembly contains overcurrent devices that are listed for continuous operation at 100% of their rating, these overcurrent devices may be loaded to 100% of their rating.

For adjustable electronic overcurrent protective devices, where the adjusting means is external to the overcurrent device, the rating of the protection is considered to be the maximum setting possible.

If the adjusting means of the overcurrent device is protected by removable and sealable covers, bolted equipment enclosure doors, or locked doors that are accessible to qualified persons only, the rating of the overcurrent device is considered to be the setting of the adjustment (long-time pickup setting). (690.8(B)(3)), (240.6(B),(C)).

690.8(A) states that the maximum calculated current is the sum of the parallel-connected PV module rated short-circuit currents multiplied by 125%

## Example

10 series connected DC modules with a short-circuit current rating of 8.9 amperes, each.
8.9 amperes × 1.25 (125%) = 11.125 amperes.
The maximum PV Source Circuit Current is 11.125 amperes (690.8(A)(1)).

The overcurrent device rating will be 1.25 (125%) times 11.125 amperes, or 13.90 amperes to satisfy 690.9(B).

When the 125% value of 690.8(A)(1) is applied to calculate the maximum PV Source Circuit current, and this is combined with the overcurrent device rating of 125% from 690.9(B)(1), the result is 156% (1.25 × 1.25 = 1.5625). 8.9 amperes × 1.5625 =13.90 amperes

Section 690.9(A) Exception states that Overcurrent Protection is not required for PV Source Circuits, if the short-circuit currents from all sources do not exceed the ampacity of the DC circuit conductors.

So, if the calculated short-circuit current from 690.8(A) and 690.9(B) is 13.90 amperes, the equipment grounding conductor may be No. 14 copper (2.08mm$^2$) from Table 250.122. This applies whether, or not, the source circuits are provided with overcurrent protection.

The equipment grounding conductors for the PV Source Circuits and PV Output Circuits (the circuit(s) from the PV combiner to the Inverter) do not have to be increased in size to compensate for voltage-drop (690.45). These circuits are DC, and are not subject to inductive reactance, as would be the case for AC circuits. The minimum size is No.14 (2.08mm$^2$).

For PV systems with a generating capacity of 100 kW or more, the PV Source Circuits may be determined through calculations made by a licensed professional electrical engineer. An industry standard that may be used to determine the maximum current of a PV system is 'Sand 2004-3535', from Sandia National Laboratories. The use of this calculation method will typically result in a lower current value than the calculation method of 690.8(A)(1)(1). However, the calculation from this standard must be not less than 70% of the value from 690.8(A)(1)(1).

In addition, for grounded DC systems, the Direct-Current System Bonding Jumper, which is used to connect the grounded conductor to the equipment grounding conductor(s), is sized in accordance with the size of the required system grounding electrode conductor (250.166). This DC System Bonding Jumper serves the same purpose as the 'Main Bonding Jumper' for AC systems. It may be installed at the source or at the first system disconnecting means supplied from the source. In this regard, it is the same as a typical Separately-Derived AC System from Section 250.30(A)(1).

Any exposed noncurrent carrying metal parts of PV module frames or panels and electric equipment enclosures must be grounded, regardless of voltage (690.43).

Metallic mounting structures (not building steel) may be <u>identified as equipment grounding</u> conductors. And <u>identified</u> bonding jumpers or devices may be used to provide an acceptable bonding means between separate metallic sections. Equipment mounted on these structures (modules) may be bonded with <u>listed devices identified</u> for this purpose. The structure must be bonded to the equipment grounding system (690.43(A),(B)).

Where equipment grounding conductors are installed for the PV arrays and the metal structure for their support, they must be contained within the same raceway, cable, or otherwise run with the PV Array circuit conductors (690.43(C)).

For PV modules, the provisions of 250.120(C) and 250.134(B), Exception No. 2 may apply. In this case, the equipment grounding conductor may not be routed with circuit conductors. And where the equipment grounding conductors are smaller than No. 6 AWG (13.3 mm$^2$), physical protection, in the form of a raceway or cable armor is required, unless the conductor is installed in such a way that physical protection is not necessary.

This information is another example of the Code Arrangement of Section 90.3, where Chapters 1-4 apply, and these general requirements may be supplemented or modified by information in Chapter 5, 6, and 7. Section 300.3(B) requires that conductors of the same circuit, including equipment grounding conductors, must be contained in the same raceway, cable, auxiliary gutter, cable tray, cablebus assembly, cable, or cord in order to limit overall circuit impedance. Section 300.5(I) identifies the same requirement for underground systems.

But, once again, the equipment grounding conductors from the PV modules are not subject to AC impedance, only DC resistance. So, installing these conductors as single conductors, with physical protection as necessary, is not a problem.

But, 690.43(C) specifically states that the equipment grounding conductors for the PV arrays are to be within the same raceway, cable, or otherwise run with the PV array circuit conductors. So 250.120(C) and 250.134(B), Exception No.2 do not apply in this case.

## Grounding Electrode System

The grounding of AC modules, or the grounding of the AC system from an Inverter to a grounding electrode (system) will be in accordance with 250.50 through 250.60. Any ferrous metallic enclosure for a grounding electrode conductor must be bonded (on both ends) to the internal grounding electrode conductor to comply with 250.64(E)(1).

For solidly grounded systems, it is common to have photovoltaic systems with both DC and AC grounding requirements. The DC grounding system must be bonded to the AC grounding system in order to limit voltage differences between these systems. This is a similar requirement to 250.50 and 250.58, and for the same reason. If two grounding electrode conductors are installed, one for the DC system and one for the AC system, a bonding conductor will be installed between these systems. The bonding conductor size is based on the larger size of the AC grounding electrode conductor or the DC grounding electrode conductor (250.66, 250.166).

Or, a DC grounding electrode conductor, sized in accordance with 250.166 may be installed from the DC grounding electrode connection point to the AC grounding electrode.

It is also common on PV systems to have an isolation transformer installed in order to separate the DC grounded circuit conductor from the AC grounded circuit conductor. Once again, there must be a bonding connection between these systems. Or, these systems may be connected to a common grounding electrode in order to establish the same zero-volts potential reference to the earth.

Any auxiliary grounding electrode (Section 250.54) installed at ground level for roof-mounted or pole-mounted arrays must be connected to the array frame or supporting structure. It may be possible to bond the frame of a roof-mounted array to the metal frame of a building, if the metal building frame is part of the grounding electrode system (250.52(A)(2)), (690.47(B)).

It should be noted that the continuity of the equipment grounding system must be maintained, even if equipment is removed for repair or replacement. The PV source and PV output circuits remain energized as long as the PV modules are exposed to light. For example, if the Inverter is removed for service, a bonding conductor must be installed to maintain the connection to the grounding system.

In addition, if the removal of equipment causes the Main Bonding Jumper in the Inverter to be disconnected, a bonding jumper must be installed to assure the grounding connection to the system grounded conductor. There may be a significant voltage-rise on the grounded conductor if the connection to the grounding electrode is interrupted.

A 'Functional Grounded PV System' has an electrical reference to ground through an element (fuse, circuit breaker, or electronic device) that is a part of a listed ground-fault protection system, and that is not solidly grounded. These systems will be at earth potential under normal conditions, but may be at an elevated voltage above earth potential during fault conditions.

## Grounding and Equipotential Bonding Agricultural Buildings (Article 547)

The consideration in this environment is excessive dust, especially where the dust combines with water. Conducting paths are easily established through this combination. And, these conducting paths may be a source of voltage differences that are harmful to livestock.

Proper grounding techniques and equipotential bonding methods are critical in controlling voltage differences due to the sensitivity of livestock to even small potential differences.

In addition, this is typically a corrosive environment due to animal excrement, especially where the excrement combines with water. These conditions may be damaging to electrical equipment and the various wiring methods that are considered to be suitable for the environmental conditions.

Wiring methods include Types UF, NMC, SE cables (copper only), covered MC cable, Rigid Nonmetallic Conduit, Liquidtight Flexible Nonmetallic Conduit, or other wiring methods that are suitable to the authority having jurisdiction.

Depending on the severity of the dust environment, the wiring methods of Section 502.10 may apply. This may include threaded rigid or intermediate metal conduit, and MI cable with listed termination fittings. Flexible connections include Liquidtight Flexible Nonmetallic Conduit. Listed fittings for these flexible conduits are required. Extra-hard usage flexible cords with dustight cord connectors, and installed in accordance with 502.10(A),(2),(5), is acceptable.

It is common in these locations to have a distribution point where overhead or underground supply conductors are extended to buildings and structures on the same premises. There will be a Site-Isolating Device, which will be pole-mounted at a height that complies with Section 230.24. It is at this location that the grounded (neutral) conductor is connected to the grounding electrode (system). And, this is the location of the Service Point, 'the point of connection between the facilities of the serving utility and the premises wiring'. So, the premises wiring begins on the load side of the service point. The site-isolating device may, or may not, be provided with overcurrent protection for the building(s) or structure(s) supply conductors. The site-isolating device must have a readily accessible operating handle. And it must be durably and legibly marked to identify it as a site-isolating device at the operating handle, or adjacent to this handle (547.9(A),(10)).

In the case of the supply conductors for agricultural buildings, the location of the Main or Supply-Side Bonding Jumper, between the grounded conductor (neutral) and the equipment grounding conductor, as well as the grounding electrode conductor connection to the grounding electrode (system), is made at the site-isolating device. And, as we have said previously, this is the location of the Service Point. And, as such, this is where the Service Conductors normally originate. And, as in other types of installations, the ungrounded conductor(s) and the grounded (neutral) conductor would extend, either overhead, or underground to the service equipment (panelboard, disconnect, etc). At the service equipment, there would be a Main Bonding Jumper between the grounded (neutral) conductor and the equipment grounding system. From this connection point, a grounding electrode conductor is installed and connected to the grounding electrode (system).

However, in this case, the Main Bonding Jumper has already been installed at the site-isolating device. And the supply conductors to the building(s) or structure(s) will include an insulated or covered equipment grounding conductor. And because the concern for agricultural buildings is even slight potential differences, and the effects of 'stray voltage' on livestock, it would be improper to bond the grounded (neutral) conductor to the equipment grounding system at the building(s) or structure(s). Such a connection would produce a voltage-rise, above earth potential, on the equipment grounding system. This is due to the current returning to the source through the grounded (neutral) conductor. In fact, it is a sound recommendation to supply loads line-to-line and not line-to-neutral to reduce neutral-to-ground voltages, which may be caused by capacitive coupling between the neutral conductor and ground. Keep in mind that we are considering even slight voltage differences that would not be detected by people. But may be a significant problem for livestock. In addition, limiting voltage-drop to a low level, possibly to as low as one percent, will be beneficial.

The equipment grounding conductor, run with the supply conductors, and connected to the grounding electrode at the building or structure will serve as the bonding means between this grounding electrode and the grounding electrode at the site-isolating device. With these grounding electrodes in parallel, the resistance-to-ground of the grounding electrode system will be reduced. This will serve to reduce the voltage-rise on systems and equipment, as well as to hold the systems and equipment at very close to earth potential.

## Equipment Bonding

We are aware that even small voltage differences in certain environments are of major concern. Certainly, agricultural buildings and adjacent areas for livestock, as well as those that are part of substations and switchyards, patient-care areas of health-care facilities, information technology equipment rooms, and areas in close proximity to swimming pools, spas and hot tubs, are examples of potential problem areas.

Section 547.10 covers 'Equipotential Planes and Bonding of Equipotential Planes'.

For indoor areas that have concrete floors, and where there is metallic equipment that may become energized, and is accessible to livestock, an equipotential plane must be installed. Fabricated bonding mesh is available for this purpose. The equipotential plane is bonded to the building or structure electrical grounding system by a solid copper conductor not smaller than No. 8 AWG (8.37 mm²). This conductor may be insulated, covered, or bare. But, due to the effects of corrosion, using an insulated conductor is recommended. The connections should be made using the appropriate fittings and connectors that are properly listed in accordance with Section 250.8(A).

For outdoor locations, the equipotential plane (mesh) will be encased in a concrete slab, where metallic equipment is located and may become energized. Of course, the equipotential plane must extend to cover the entire area where the livestock will stand. And, this equipotential plane will be bonded to the equipment grounding system.

An important consideration is to recognize that when stepping away from the equipotential plane there may be a voltage difference, while not detected by people, it may have an adverse effect on livestock.

## Swimming Pools. Fountains, and Similar Installations-Grounding and Bonding

This sensitive electrical environment is commonly misunderstood and not provided with the appropriate wiring methods that will assure a safe installation.

So, let's begin with the grounding requirements of 680.6. This Section includes a list of equipment that requires grounding.

1. Through wall lighting assemblies and underwater luminaires
2. Electrical equipment that is located within 5 feet (1.5 meters) of the inside wall of the pool, hot tub, spa, or fountain
3. The electrical equipment that is associated with the recirculating system

4. Junction boxes
5. Transformer and power supply enclosures
6. Ground-fault circuit interrupters
7. Panelboards (subpanels) that are not part of the service equipment

Fixed or stationary equipment, such as a filter pump, is often connected by a flexible cord to facilitate removal for maintenance, or to protect this equipment from the effects of the weather. The cord is limited to a 3 foot (900mm) length. This cord must have a copper equipment grounding conductor not smaller than No. 12 AWG (3.30mm²),(680.8).

It should be noted that a flexible cord must have a length of 25 feet for the recirculating pump motor for a storable pool. This 25 foot length is a UL requirement, and it is meant to prevent the use of extension cords in the pool area. The flexible cord is required to have an equipment grounding conductor. The filter pump is double insulated (or equivalent), and the equipment grounding conductor is used to ground the internal noncurrent carrying metal parts of the equipment. The flexible cord is required to have an integral ground-fault circuit interrupter located within 12 inches (300 mm) of the attachment plug (680.30, 680.31).

A storable pool is defined as having a maximum water depth of 42 inches (1.0 mm). The listed recirculating pump motor for a storable pool is not meant to be used for permanently installed pools (680.2).

For a storable pool, storable hot tub, or storable spa, receptacles must be not less than 6 feet (1.83m) from the inside wall of this equipment (680.34). This receptacle(s), and any other receptacles within 20 feet (6.0m) of this equipment must be GFCI protected (680.32).

For hot tubs and spas, the recirculating pump may have a flexible cord up to 15 feet (4.6m) in length where protected by a GFCI (680.42(A)(2)).

For cord-and-plug connected equipment associated with fountains, the cord must have an insulated copper conductor and the supplied equipment must be GFCI protected. This flexible cord is either exposed to, or immersed in water. It must be suitable for extra-hard usage and be listed with a 'W' suffix (680.55, 680.56).

For cord-and-plug connected equipment, and any fixed equipment associated with fountains, GFCI protection is required (680.56(A)).

## Wiring Methods

It is relatively common to extend a feeder from the service equipment to a panelboard located at the location of the pool. This feeder will have the ungrounded conductor(s), a neutral conductor, and an <u>insulated equipment</u>

grounding conductor (680.25(A)). These feeder conductors may be copper, aluminum, or copper-clad aluminum if installed in a noncorrosive environment. However, branch circuit wiring, including equipment grounding conductors, must be copper.

The feeder wiring methods include:

1. Liquidtight flexible nonmetallic conduit (Article 356)
2. Rigid PVC (Article 352)
3. Reinforced thermosetting resin conduit (Article 355)
4. EMT (Article 358)
5. ENT (Article 362)
6. MC cable (Article 330)
7. Rigid metal conduit (Article 344)
8. Intermediate metal conduit (Article 342)
9. AC cable (Article 320)
10. NM, NMC, NMS cable (Article 334)
11. Flexible metal conduit (Article 348)

Branch circuit wiring in a noncorrosive environment includes any of the wiring methods referenced in Chapter 3. And, in corrosive environments, as described in 680.14(A), rigid metal conduit, intermediate metal conduit, rigid polyvinyl chloride conduit, and reinforced thermosetting resin conduit are acceptable.

Section (680.14(A)) describes a corrosive environment as locations where pool sanitation chemicals are stored, as well as areas with circulation pumps, automatic chlorinators, filters, open areas beneath decks that are adjacent to or abutting the pool structure, and similar locations. Suitable wiring methods include rigid metal conduit, intermediate metal conduit, rigid PVC, and reinforced thermosetting resin conduit.

## Equipotential Bonding

Once again, the concept of reducing voltage differences in these areas is of paramount importance. Great care must be exercised to keep any metallic surfaces that are associated with a pool at the same potential. This includes conductive pool shells, structural reinforcing steel, the forming shells of underwater luminaires, metal ladders, metal diving board stands, reinforcing steel or mesh in concrete walkways around the pool, and the recirculating pump equipment.

This reverts back to the definition of the term 'Bonded' or 'Bonding' in Article 100, 'connected to establish electrical continuity and conductivity'.

The conductor that is used to make connections is a solid copper conductor, insulated, covered, or bare, and not smaller than No. 8 AWG (8.37 mm$^2$) (680.26(B)).

The use of a conductive pool shell, an unencapsulated structural reinforcing steel forming shell, or a copper conductor grid, may be used as a common bonding reference to which all metallic equipment is connected (680.26(B)(1)).

Connections to the common grid must comply with 250.8(A), that is 'listed' mechanical connectors or exothermic connections.

It should be noted that where structural reinforcing steel is not available, or, where the reinforcing steel is encapsulated in a nonconductive compound, a solid No. 8 AWG (8.37 mm$^2$) bare solid copper conductor (or larger) may be used as a common bonding means. This bonding conductor must follow the contour of the perimeter of the pool. It must be 18 inches (450 mm) to 24 inches (600 mm) from the inside wall of the pool, and it must be secured within or under the perimeter surface 4 to 6 inches (100 –150 mm) below the sub grade (such as a concrete walkway).

It should also be noted that the bonding conductor serves the purpose of maintaining electrical continuity and limiting voltage differences between any conducting surface associated with the pool. It is not a grounding conductor. There is no need to connect this conductor to the equipment grounding system in a subpanel, or to extend this bonding conductor to the service equipment, or to connect the bonding conductor to a grounding electrode. Connections, such as these, serve no useful purpose in enhancing the bonding system integrity.

And, it is not acceptable to use the bonding grid as a grounding electrode (250.52(B)(3)). As we have stated, the bonding grid is provided to equalize voltage differences. To use this bonding grid as a grounding electrode may introduce dangerous voltages to the bonding system.

The bonding requirements for permanent pools do not apply to storable pools.

However, they do apply to outdoor spas and hot tubs, as well as pools and tubs for therapeutic use, and hydromassage bathtubs. (680.42(B), 680.62(B), 680.74(B)).

To clarify the equipotential bonding requirements of 680.26(B)(2) for hot tubs or spas installed outdoors, as they apply to 'Perimeter Surfaces', the following conditions negate the need for equipment bonding around the perimeter of the hot tub or spa:

1. The spa or hot tub must be listed, labeled, and identified as a self-contained spa for above ground use.
2. The spa or hot tub is not identified for only indoor use.
3. The installation complies with the manufacturer's instructions and is installed on or above grade.
4. The top rim of the hot tub or spa is at least 28 inches (710 mm) above all perimeter surfaces that are within 30 inches (760 mm) horizontally

from the hot tub or spa. The listing of this equipment is in accordance with ANSI/UL 1563-2010.

However, for other than the waiver of equipotential bonding for perimeter surfaces (680.26(B)(2)), the other bonding requirements of 680.26 apply.

Permanently installed pools for therapeutic use that are inground, or aboveground, must comply with the equipotential bonding requirements of 680.26 (680.61).

For therapeutic tubs that are used for submersion and treatment of patients, the following parts must be bonded (680.62(B)):

1. All metal fittings within or attached to the tub structure
2. Metal parts that are associated with the tub water recirculating equipment, including pump motors
3. Any metallic cables or metal piping that are within 5 feet (1.5 m) from the inside walls of the tub
4. Any metal surface within 5 feet (1.5 m) from the inside walls of the tub, and not separated by a permanent barrier.
5. Any electrical device or control that is not a part of the tub and located within 5 feet (1.5 m) from the tub.

Section 680.62(C) covers the methods of bonding of metal parts. The most familiar, and probably the best method of assuring bounding integrity, is through the use of a solid copper conductor, not smaller than No. 8 AWG ($8.37mm^2$), which may be insulated, covered, or bare.

For Hydromassage Bathtubs, the provisions of 680.74(B) apply. Once again, a solid copper conductor, No. 8 AWG ($8.37 mm^2$), insulated, covered, or bare must be used to bond all metal parts that are associated with the circulating water, including the pump motor. However, this does not apply to a double-insulated pump motor because such a bonding connection may compromise the system of double insulation. But the No. 8 AWG copper bonding conductor is required to be run to the location of the pump motor with sufficient length to accommodate a replacement pump motor, which may not be of the double-insulated type.

# Article 517

## Health Care Facilities-Grounding and Bonding

In what type of environment is proper bonding and grounding more important than in a health care facility, especially in patient-care areas?

In addition, health care facilities are required to have two independent sources of power (517.35). One of these power sources may be a utility service, possibly even more than one utility service. Ideally, but not always available, a second utility supply, which is derived from a separate distribution system.

And the second of these supply systems may be a generator driven by a prime mover, which is located on the premises, or two generating units on the premises, one for normal power, and the second as a backup unit. Or, a battery backup system which may be recognized as a source of supply.

The most common sources of power for health care facilities are a utility supply and an onsite generator, (700.12), (NFPA 110 - 2013).

Obviously, today there are quite sophisticated types of medical equipment in health care facilities that are susceptible to all types of disturbances, whether internally, or from outside sources.

As a normal course of action, the utility supply system will be connected to a grounding electrode at the supply transformer, and there will be another connection to ground at the Service Equipment (250.24).

And, there will be another connection to a grounding electrode (system) at the alternate power source (generator) (250.30), if the generator is a separately-derived system.

It would also seem routine that both of these systems are connected to the same grounding electrode (system). And, hopefully, this is the case.

This is critical in keeping a constant potential reference to the earth ('0' volts, or as close to '0' volts as possible) between the normal and the backup system.

Where the grounded conductors (neutrals) of both systems are properly interconnected, ground-potential shift will not cause a disturbance to equipment supplied from the two power sources.

Where the power sources are relatively close, each system (service equipment), (generator), may be connected to the grounding electrode (system) by grounding electrode conductors that are of relatively equal length.

But if these power sources are not close to each other, it would be beneficial to provide a solid connection between the two system neutral points, with only one of these neutral points connected to the grounding electrode (system).

With both system neutral points properly bonded and only one grounding electrode conductor extended to the grounding electrode (system), there cannot be a ground potential shift.

In addition, where the separately-derived system is outside of the building or structure, a connection to a grounding electrode is to be made at this location (250.30(C)). This grounding electrode must be bonded to any other grounding electrodes that are present at the building or structure. This connection is established through the supply neutral conductor. Just as in the case of a service supply system (250.24(A)(2)), where the transformer is outside of the building or structure, the concern is related to the effects of lightning, transient voltages from the serving utility, and possible contact with other conductors.

## Patient Care Space

The consideration of proper bonding and grounding cannot be of greater importance than at patient care areas of health care facilities, where, even very small voltage differences may be harmful or fatal to people.

I remember from many years (even decades) ago, where various protection techniques were investigated for Information Technology Equipment Systems. And, consideration of the effects of static electricity was an important concern. A statement was made that, as a person moves across a floor, static charges would form on the individual as layers of clothing contact and separate, and as the person's shoes make contact and separate from the floor. The static electricity may be on the order of 2000-3000 volts, which is below the threshold of feeling of most people. However, if this voltage difference is coupled into sensitive equipment, it may be the equivalent of a miniature lightning bolt for the equipment and cause data corruption, and possible damage. There was a recommendation that the individual approaching this equipment, pause for a minute, or so, before making contact. This sounded like a logical approach to avoid equipment damage.

Of course, there are many ways to mitigate these unwanted voltage differences. Humidity, conductive footwear and floors, and conductive clothing, etc., are effective methods of controlling these conditions.

Section 517.2 defines a 'Patient Bed Location', 'Patient Care Space', and 'Patient Care Vicinity'.

The 'Patient Care Vicinity' is defined as, 'A space, within a location intended for the examination and treatment of patients, extending 6 feet (1.8 m) beyond the normal location of the patient bed, chair, table, treadmill, or other device that supports the patient during examination and treatment and extending vertically 7 feet 6 inches (2.3 m) above the floor'.

Certainly, this is a critical environment, where, as we have previously stated, even very low voltage differences and current levels may pose a significant threat to patients, especially those whose body resistance has been compromised by a medical procedure.

A 'Patient Equipment Grounding Point' is defined as a jack or terminal that serves as the collection point for redundant grounding of electrical appliances serving a patient care vicinity, or for grounding other items in order to eliminate electromagnetic interference problems.

Section 517.19(D) permits, but does require, a 'Patient Equipment Grounding Point' in a 'Patient Care Vicinity'. The grounding terminals of all grounding-type receptacles would be bonded to the 'Patient Equipment-Grounding Point' by a bonding jumper not smaller than No. 10 AWG (5.26mm$^2$).

The 'Patient Equipment Grounding Point' may be arranged as a busbar in the 'Patient Care Vicinity'. And it would serve to keep the equipment grounding system in this area at, virtually, the same potential.

It should be noted that the length of the bonding jumpers should be limited to the common bonding point. Conversely, if a Patient Equipment Grounding point is not provided in the patient care vicinity, a reference grounding point should be in close proximity to the patient care vicinity in order to reduce voltage differences.

The bonding of the equipment grounding terminal bus of the normal and essential branch-circuit panelboards, where they serve the same patient care vicinity is required in accordance with 517.14. The normal branch-circuit panelboard is supplied by the normal power source (normally, the serving utility). And the essential branch-circuit panelboard is supplied by the essential electrical system (alternate power source). It makes sense that where these systems supply the same patient care vicinity that the proper bonding of each panelboard equipment ground is accomplished with the required minimum No.10 AWG (5.26mm$^2$) insulated, continuous copper conductor.

As another example of Section 90.3, and an amendment of a general-rule from Chapter 2, specifically 250.118, Section 517.13(A) requires that branch circuits serving patient care areas be provided with an effective ground-fault current path. While metal raceways, metallic cable assemblies, or metallic cable sheath may be acceptable as an equipment grounding conductor (250.118), their use as the sole equipment grounding path in a patient care area is not acceptable.

These metallic wiring methods must be supplemented with an insulated copper equipment grounding conductor (517.13(B)).

The insulated copper equipment grounding conductor is sized in accordance with 250.122. And it is connected to the grounding terminals of all receptacles, and all metal surfaces of fixed equipment that may become energized where the voltage exceeds 100 volts, except the metal frames of luminaires that are more than 7 feet, 6 inches (2.3m) above the floor. Metal faceplates for switches and receptacles are considered effectively grounded by their attachment to a grounded box or device.

Isolated ground receptacles (250.146(D)), (406.3(D)) are not permitted in a patient care vicinity. The grounding terminal of this type of receptacle is not connected to the receptacle mounting yoke (strap). The redundant grounding path provided by the metal raceway or metallic cable assembly, in conjunction with the internal insulated copper equipment grounding conductor, as required by 517.13(A)(1), is not provided by the use of these types of receptacles (517.16 (A)). However, isolated-ground receptacles are acceptable outside of a patient care vicinity in accordance with 517.16(B)(1),(2).).

Ground-fault circuit interrupters are required in bathrooms of nondwellings (210.8(B)(1)), and this includes the receptacles(s) installed in bathrooms in critical care areas of health care facilities. However, where a toilet and sink are installed within the patient room, and not in an adjacent room, a receptacle adjacent to the sink is not required to have this protection. The reasoning for this exemption is that this receptacle may be used for equipment serving the patient and the inadvertent tripping of the GFCI may pose a problem for the patient (517.21).

Section 517.160 covers the use of Isolated Power Systems, which were commonly used in anesthetizing locations that were considered to be Class I Hazardous (Classified) Locations. This is no longer the case in the United States, as flammable anesthetics are not used (Ethylene, Ethyl Ether, Cyclopropane, etc.). But, flammable anesthetics may be used at some time in the future. However, while working in certain hospitals in Africa over the past several years, I have found that the use of flammable anesthetics is relatively common.

Isolated Power Systems are commonly supplied by an isolation transformer(s) or generator. The secondary circuit of an isolation transformer(s) is ungrounded and the maximum voltage is 600 volts. Systems supplied from a generator(s) or another type of system will also be ungrounded (517.160(A)(2)).

This equipment is not permitted within the hazardous (classified) location, which extends to a distance of 5 feet (1.52 m) above the floor. Areas above this height are considered 'above a Hazardous (Classified) Location'.

An isolation transformer will have a metallic shield between the primary and secondary, which are wound on separate cores that are adjacent to each other. This electrostatic shield is to be connected to the 'reference grounding point,' which is the equipment ground bus in the panel(s) supplying the patient care area (517.160(A)(2)).

The identification of the ungrounded conductor is: No.1-Orange, with one or more colored stripes, other than white, green, or gray. No. 2-Brown, with one or more colored stripes, other than white, green, or gray. For 3-phase systems, the third conductor is Yellow, with one or more colored stripes, other than white, green, or gray.

Where 125 volt, 15 or 20 ampere, single-phase receptacles are supplied, the orange conductor is used to connect to the receptacle terminal for the grounded (neutral) conductor (517.160(A)(5)).

A Line Isolation Monitor (517.160(B) is to be installed to monitor leakage current on the ungrounded system. An alarm system (audible and visual) is designed to be energized when the leakage current to ground reaches 5 mA.

Because leakage current will be affected by the normal capacitive coupling of conductors to the equipment grounding system, circuit lengths should be limited.

It should be noted that grounding and bonding in all anesthetizing areas, whether the area has been classified as hazardous, or not, must be assured by connecting fixed electrical equipment, operating at over 10 volts, to an equipment grounding conductor. Wiring methods for anesthetizing areas that are not Hazardous (Classified) Locations, must be installed in a metal raceway or metallic cable assembly that is included in 250.118 as a type of equipment grounding conductor.

If the anesthetizing area is a Hazardous (Classified) Location, the grounding and bonding requirements of Class I locations must comply with 501.30, (517.62).

The following questions are based on references from the National Electrical Code that are directly associated with grounding and bonding. Be sure to review the appropriate Code Sections, including Exceptions and Informational Notes. This exercise will serve to reinforce your understanding of this important topic. Remember, the NEC is not intended as a design specification (90.1(A)).

As always, I welcome your comments and suggestions.

**Gregory P. Bierals**
**Electrical Design Institute**

1.  A Supply-Side Bonding Jumper is installed on the supply side of a service e.g., meter enclosure, or within a service equipment enclosure, or a System Bonding Jumper for a separately derived system, and is sized in accordance with _____.
    A) Table 250.66
    B) Table 250.122
    C) Table 250.102(C)(1)
    D) Table 310.16(B)(16)

2.  Grounded electrical systems are connected to the earth in a manner that will limit the voltage imposed by lightning, line surges, or unintentional contact with higher voltage lines, and that will _____the voltage to earth during normal operation.
    A) reduce
    B) stabilize
    C) increase
    D) limit

3.  An effective ground-fault current path is a    path which extends from any point on the wiring system where a _____may occur to the electrical supply source.
    A) conducting-short- circuit
    B) high impedance-ground-fault
    C) equipment ground-overcurrent
    D) low impedance-ground-fault

4.  A ground-fault detection system is required for a high-impedance grounded system?
    A) True
    B) False

5.  _____ current, caused by downstream connections between the grounded (neutral) conductor and the equipment grounding system may affect sensitive electronic equipment.
    A) Circulating
    B) DC
    C) Objectionable
    D) Temporary

6.  Class 2 load side circuits for suspended ceiling power distribution systems are required to be grounded.
    A) True
    B) False

7. Rod-type grounding electrodes smaller than 5/8" (15.87mm) must be .
   A) listed
   B) certified
   C) identified
   D) A and C

8. The structural reinforcing steel for a swimming pool may be used as a grounding electrode, or as part of the grounding electrode system.
   A) True
   B) False

9. Grounding electrode conductors smaller than _____ AWG shall be protected from physical damage by RMC, IMC, PVC, RTRC-XW, EMT, or cable armor.
   A) 6
   B) 8
   C) 10
   D) 4

10. Separate grounding electrodes, including those installed for communications systems and strike termination devices are required to be _____ to the building or structure grounding electrode system.
    A) grounded
    B) bonded
    C) isolated
    D) none of these

11. The grounding electrode conductor for an AC system consisting of ungrounded copper 3/0 AWG service-entrance conductors is _____.
    A) 4 AWG
    B) 6 AWG
    C) 1 AWG
    D) 2 AWG

12. The paralleling efficiency of ground rods increases with their spacing that is equivalent to _____ the length of the longest rod.
    A) 2 times
    B) 3 times
    C) 4 times
    D) 5 times

13. A ground-fault protection system is required for solidly grounded 3-phase, 4-wire, Wye connected systems operating at up to 1000 volts, phase-to-phase, and over 150 volts, phase-to-neutral, where the service or feeder disconnect is rated _____ amperes, or more.
A) 800
B) 1200
C) 1000
D) 600

14. The various types of equipment grounding conductors are included in _____.
A) 250.122
B) 250.118
C) 250.66
D) 250.102

15. Where the ungrounded conductor(s) is increased in size to compensate for voltage-drop, a _____ increase in size must be made to the size of the equipment grounding conductor, unless the AHJ permits the EGC size to be determined by a qualified person.
A) significant
B) proportional
C) 2 times
D) 3 times

16. Metal cable tray may serve as an equipment grounding conductor where the cable tray sections are electrically _____.
A) bonded
B) grounded
C) continuous
D) A and C

17. The bonding and grounding of a separately-derived system must be done at the source or at the _____ disconnecting means or overcurrent device.
A) first or second
B) at any single point
C) where the grounded conductor is present
D) first

18. Ground rods and plate electrodes shall be installed _____, where practicable.
    A) below the service panel
    B) at least 3 feet deep
    C) below permanent moisture level
    D) all of these

19. Bonding shall be provided where necessary to ensure electrical continuity and the capacity to conduct any _____ current likely to be imposed.
    A) High-impedance
    B) fault
    C) isolated ground
    D) arcing

20. The frame of a portable or vehicle-mounted generator shall not be required to be connected to a _____, where the generator supplies equipment mounted on the generator and/or cord-and-plug connected equipment supplied from receptacles mounted on the generator.
    A) equipment grounding conductor
    B) grounded conductors
    C) equipment grounding conductor
    D) grounding electrode

21. All branch circuits serving patient care spaces shall be provided with an effective ground-fault current path by installation in a _____ system or metallic cable that qualifies as an equipment grounding conductor in accordance with 250.118. An _____ copper equipment grounding conductor must be run with the branch circuit conductors.
    A) metal raceway-insulated
    B) metal conduit-identified
    C) cable armor –bare
    D) A and C

22. Where ground-fault protection is required for a health care facility, an additional level of ground-fault protection is required downstream on every feeder, where the rating of the feeder disconnecting means is 1000 amperes, or more.
    A) True
    B) False

23. Where photovoltaic modules are mounted on metal supports, devices for bonding the equipment to the metal supports shall be_____.
    A) listed
    B) labeled
    C) identified
    D) all of these

24. Motors that are protected by instantaneous-trip circuit breakers are permitted to be provided with equipment grounding conductors based on the equivalent size of a _____.
    A) inverse-time circuit breaker
    B) nontime-delay fuse
    C) dual-element time-delay fuse
    D) overload device

25. Ground-fault protection of equipment is _____ in a fire pump power circuit.
    A) not to be installed
    B) permitted
    C) provided
    D) acceptable

26. Equipment grounding conductors are to be run with the other circuit conductors, within the same raceway, cable, or cord. Exceptions to this requirement include _____ or _____.
    A) SE or USE cable
    B) nongrounding receptacle replacements
    C) branch circuit extensions
    D) B and C

27. On some AC circuits, the equipment grounding conductor may be larger than the ungrounded circuit conductors.
    A) True
    B) False

28. In an ungrounded system, a low-impedance path for fault current must be established from any point on the wiring system to the electrical supply source to facilitate the operation of the overcurrent devices in the event

a _____ ground-fault develops on an opposite phase of the wiring system.
A) second
B) arcing
C) low-impedance
D) bolted

29. _____ are not permitted to be installed within a patient care vicinity.
A) ground-fault circuit interrupters
B) isolated grounding receptacles
C) arc-fault circuit interrupters
D) ungrounded receptacles

30. Double-locknut protection is a permitted bonding means in Class I, Division 1 and 2 Hazardous (Classified) Locations.
A) True
B) False

31. Intrinsically safe apparatus, enclosures, and raceways, if of metal, _____ be connected to the equipment grounding conductor.
A) shall
B) shall not
C) are not required to
D) may

32. An auxiliary grounding electrode, used as a supplement to the equipment grounding system, _____ be required to be bonded to the building grounding electrode system.
A) shall
B) shall not
C) must
D) A and C

33. Equipment bonding jumpers on the load side of an overcurrent device may be outside of a raceway for lengths not to exceed _____ feet.
A) 8
B) 6
C) 3
D) 10

34. Where signal reference grid structures are installed in an Information Technology Equipment room for the purpose of creating an equipotential plane at higher frequencies, this grid _____ bonded to the equipment grounding system for the Information Technology Equipment.
    A) shall be
    B) shall not
    C) may
    D) may not

35. Type 2 Surge Protective Devices may be connected on the supply (line) side of the service disconnecting means in accordance with_____.
    A) 230.82(4)
    B) 285.24
    C) 230.82(8)
    D) 285.1

36. Grounding electrode conductors installed within ferrous metal raceways for physical protection in lengths of 3 feet or less, are not required to be bonded to this enclosure.
    A) True
    B) False

37. Wet-niche luminaires shall be connected to an equipment grounding conductor not smaller than _____ AWG
    A) 14
    B) 12
    C) 10
    D) 6

38. Where a grounding electrode conductor is connected to a ground ring, and then extended to a ground rod, the minimum size copper bonding conductor to the rod is _____ AWG.
    A) 6
    B) 4
    C) 8
    D) 2

39. A copper system bonding jumper for 2-500 kcmil copper conductors in parallel is _____ AWG.
    A) 6
    B) 1
    C) 2
    D) 2/0

40. A copper grounding electrode conductor for an antenna mast shall not be smaller than _____ AWG.
    A) 10
    B) 8
    C) 6
    D) 12

41. A Functional Grounded Photovoltaic System has an electrical reference to ground that is solidly grounded.
    A) True
    B) False

42. Metal fences enclosing, and other metal structures in or surrounding a substation, shall be grounded and bonded to limit step, touch, and _____ voltages.
    A) ground
    B) potential
    C) transfer
    D) all of these

43. In order to limit increased voltages associated with a lightning protection system, the system ground terminals shall be isolated from the building or structure grounding electrode system.
    A) True
    B) False

44. A _____ or _____ type receptacle may be used as a replacement for a nongrounding type receptacle where an equipment grounding conductor does not exist in the receptacle enclosure.
    A) AFCI
    B) GFCI
    C) nongrounding
    D) B and C

45. The conductors between a surge arrester and the line and the surge arrester grounding connection shall not be smaller than _____ AWG copper or aluminum.
    A) 8
    B) 6
    C) 4
    D) 2

46. For direct-current systems that are grounded, the size of the bonding jumper between the grounded conductor and the equipment grounding system shall not be smaller than the grounding electrode conductor from 250.166, and shall comply with _____.
   A) 250.66
   B) 250.168
   C) 250.122
   D) 250.28(A),(B), and (C)

47. Grounding electrode conductors should not be any longer than necessary to complete the connection and so that unnecessary_____ and _____ are avoided.
   A) obstructions
   B) bends
   C) loops
   D) B and C

48. The reinforcing members in the footings of existing buildings or structures shall not be required to be a part of the grounding electrode system where the reinforcing bars or rods are not accessible without disturbing the concrete.
   A) True
   B) False

49. For a high-impedance grounded neutral system, the grounded system conductor shall have an ampacity of not less than the maximum current rating of the grounding impedance, and in no case shall this conductor be smaller than _____ AWG copper, or _____ AWG aluminum or copper-clad aluminum.
   A) 8 – 6
   B) 4 – 2
   C) 10 – 8
   D) 3 – 2

50. The primary pad base plate of the wireless power transfer equipment of an electric vehicle shall be grounded or a part of a _____ system.
   A) bonded
   B) DC
   C) isolated
   D) double insulation

# Answer Key

| | | |
|---|---|---|
| 1. | C | Article 100-Definition-Bonding Jumper, System 250.2-Definition-Bonding Jumper, Supply-Side - Table 250.102(C)(1) |
| 2. | B | 250.4(A)(1) |
| 3. | D | 250.4(A)(5), 250.4(B)(4) |
| 4. | A | 250.4(A)(5), 250.36(2), 250.187(2) |
| 5. | C | 250.6(A),(B),(C),(D) |
| 6. | B | 250.22(6), 393.60(B) |
| 7. | A | 250.52(A),(5),(b) |
| 8. | B | 250.52(B),(3) |
| 9. | A | 250.64(B),(3) |
| 10. | B | 250.50, 250.53(B), 250.58, 250.60, 250.106, 800.100(A),(B), 810.21(J), 820.100(D), and 830.100(D) |
| 11. | A | Table 250.66 |
| 12. | A | 250.53(3), Informational Note |
| 13. | C | 215.10, 230.95, 240.13 |
| 14. | B | 250.118 (1 through 14) |
| 15. | B | 250.122(B) |
| 16. | D | 250.118(11), 392.18, 392.60 |
| 17. | D | 250.30(A),(1) |
| 18. | C | 250.53(A),(1) |
| 19. | B | 250.90 |
| 20. | D | 250.34(A),(B) |
| 21. | A | 517.13(A),(B) |
| 22. | B | 517.17(B) requires GFPE for every downstream feeder, regardless of the rating of the disconnecting means. |
| 23. | D | 690.43(B) |
| 24. | C | 250.122(D),(2) |
| 25. | A | 695.6(G) |
| 26. | D | 250.130(C), 300.3(B), 690.43(C) |
| 27. | B | 250.122(A) |
| 28. | A | 250.4(B),(4) |
| 29. | B | 517.16(A) |
| 30. | B | 501.30(A) |
| 31. | A | 504.50(A) |
| 32. | B | 250.54 |
| 33. | B | 250.102(E),(2) |
| 34. | A | 645.15 |
| 35. | C | 285.24(A) |

36.  B   250.64(E),(1)
37.  B   680.23(F),(2)
38.  D   250.66(A), 250.64(F)
39.  D   250.102(C),(1)
40.  A   810.21(H)
41.  B   690.2 – Definition-Functional Grounding PV System
42.  C   250.194
43.  B   250.106, 250.60, 250.50, 250.58
44.  D   406.4(D),(2),(b),(c)
45.  B   280.23
46.  D   250.168
47.  D   250.4(A),(1), Informational Note No.1
48.  A   250.50, Exception
49.  A   250.36(B)
50.  D   625.101

# Index

# About the Author

It started on August 3, 1964 as I started to work in the electrical trade through the auspices of Local 52, IBEW in Newark, N.J. Seven months later, I became a member of this Local, which was merged with Local 164 (Paramus, N.J.) in 2000.

I served my apprenticeship until March 1, 1969, at which time I became a Journeyman-Wireman. I gained a wealth of knowledge and experience during this period and the years that followed.

In March of 1978, I developed an association with a training/consulting company in Trenton, N.J. This company had a request from a client in Philadelphia to present an electrical training course for their maintenance personnel. The class was scheduled for four weeks. I was asked to conduct this course, and despite not having any teaching experience, I decided to become an instructor. Fortunately, the class went well, and the training was extended for another four weeks. In the meantime, this training/consulting company offered me a permanent position.

In September of 1978, I started to offer courses on the topic of the National Electrical Code. The key was to find a method of instruction that would benefit the students by keeping their attention during the three day course period. And, to foster an interest in this complex document that would serve them well beyond our brief time together. In later years, I developed and presented courses on the topics of Grounding Electrical Distribution Systems, Designing Overcurrent Protection, Electrical Systems In Hazardous (Classified) Locations, and Electrical Equipment Maintenance. These courses were offered by my company, Electrical Design Institute, and several universities, including the University of Wisconsin, George Washington University, North Carolina State University, the University of Toledo, and the University of Alabama.

In 2021, I authored books entitled, The NEC and You, Perfect Together, Grounding Electrical Distribution Systems, and Designing Overcurrent Protection, NEC Article 240 and Beyond. These books are published by River Publishers.

*Gregory P. Bierals*
May 24, 2021

For Product Safety Concerns and Information please contact our
EU representative GPSR@taylorandfrancis.com Taylor & Francis
Verlag GmbH, Kaufingerstraße 24, 80331 München, Germany